T0406823

Palgrave Studies in Animals and Literature

Various academic disciplines can now be found in the process of executing an 'animal turn', questioning the ethical and philosophical grounds of human exceptionalism by taking seriously the non-human animal presences that haunt the margins of history, anthropology, philosophy, sociology and literary studies. Such work is characterised by a series of broad, cross-disciplinary questions. How might we rethink and problematise the separation of the human from other animals? What are the ethical and political stakes of our relationships with other species? How might we locate and understand the agency of animals in human cultures?

This series publishes work that looks, specifically, at the implications of the 'animal turn' for the field of English Studies. Language is often thought of as the key marker of humanity's difference from other species; animals may have codes, calls or songs, but humans have a mode of communication of a wholly other order. The primary motivation is to muddy this assumption and to animalise the canons of English Literature by rethinking representations of animals and interspecies encounter. Whereas animals are conventionally read as objects of fable, allegory or metaphor (and as signs of specifically human concerns), this series significantly extends the new insights of interdisciplinary animal studies by tracing the engagement of such figuration with the material lives of animals. It examines textual cultures as variously embodying a debt to or an intimacy with animals and advances understanding of how the aesthetic engagements of literary arts have always done more than simply illustrate natural history. We publish studies of the representation of animals in literary texts from the Middle Ages to the present and with reference to the discipline's key thematic concerns, genres and critical methods. The series focuses on literary prose and poetry, while also accommodating related discussion of the full range of materials and texts and contexts (from theatre and film to fine art, journalism, the law, popular writing and other cultural ephemera) with which English studies now engages.

Ruth Hawthorn • John Miller
Editors

Animals in Detective Fiction

Editors
Ruth Hawthorn
University of Lincoln
Lincoln, UK

John Miller
University of Sheffield
Sheffield, UK

ISSN 2634-6338 ISSN 2634-6346 (electronic)
Palgrave Studies in Animals and Literature
ISBN 978-3-031-09243-5 ISBN 978-3-031-09241-1 (eBook)
https://doi.org/10.1007/978-3-031-09241-1

Cover illustration: GRANGER - Historical Picture Archive / Alamy Stock Photo

This Palgrave Macmillan imprint is published by the registered company Springer Nature Switzerland AG.
The registered company address is: Gewerbestrasse 11, 6330 Cham, Switzerland

Acknowledgements

This book has been a long time coming. We would like to thank the outstanding roster of contributors for their diligence, expertise, kindness and patience in bringing this volume to fruition. Allie Troyanos and Rachel Jacobe have been models of professionalism throughout. We are grateful to the series co-editors of Palgrave Studies in Animals and Literature, Susan McHugh and Bob McKay for their support for the project. As always we owe a great debt to colleagues and students at the University of Lincoln and the University of Sheffield for making the book possible in so many ways. Particular thanks are due to Emily Thew and Sarah Bezan, who played important roles in shaping and realising the project.

Contents

Notes on Contributors

Joseph Anderton is Senior Lecturer in English Literature at Birmingham City University. He is the author of the monograph, *Beckett's Creatures: Art of Failure after the Holocaust* (Bloomsbury, 2016) and the journal article "Vegetating Life and the Spirit of Modernism" in *Modernism/modernity* (John Hopkins, 2019). He has published several journal articles and book chapters on literary animal studies and the non-human, in relation to Samuel Beckett, Franz Kafka, J.M. Coetzee and Paul Auster, including in *Beyond the Human-Animal Divide* (Palgrave, 2017), *Twentieth-Century Literature* (Hofstra University, 2016), *Screening the Nonhuman* (Lexington, 2016), *Performance Research* (Routledge, 2015), and *Beckett and Animals* (CUP, 2013).

Nathan Ashman is Lecturer in Crime Writing at the University of East Anglia and the author of *James Ellroy and Voyeur Fiction* (2018). His research spans the fields of crime fiction, contemporary American fiction, and ecocriticism, with a particular specialism in the works of James Ellroy.

Karin Molander Danielsson holds a PhD in English from Uppsala University, Sweden. Her thesis investigated special interests and seriality in contemporary detective fiction. She has been Senior Lecturer in English at Mälardalen University, Västerås, Sweden since 2006, and a member of the research group Ecocritical forum since 2008. Her research interests are animal studies, ecocriticism, detective fiction, American literary naturalism, and narrative studies. Publications include "Unreadable Nonhumans, Ambiguity and Alterity in Eric Linklater's Short Fiction" (*Green Letters*, 2022) and "'And in that Moment I Leapt upon his Shoulder.' Non-human Intradiegetic Narrators in *The Wind on the Moon*" (*Humanities*, 2017).

She is currently writing and editing a volume on the non-human in American Literary Naturalism.

Briony Frost holds a PhD from the University of Exeter, where she was also Lecturer in Early Modern Literature. She taught for ten years at Plymouth University, where she retains a research affiliation. She is currently working for the University of Bath as a Curriculum Developer on a cross-institutional educational change project, and for OPEN Health Medical Communications as a Learning Design and Development Specialist. Her research interests include intersections between memorial, sensorial, martial and meteorological cultures, civic pageantry, representations of animals, and witchcraft in early modern literature and drama.

Alexandra Hauke is Lecturer in American Studies at the University of Passau, Germany, where her research and teaching focus on North American literatures, Indigenous studies, detective fiction, American film and TV, folk horror, and digital cultures. She has co-edited essay collections on Native American survivance as well as twenty-first-century Canadian literatures and has written on law and legal cultures in Native American detective fiction, American ecofeminist gothic fiction, blackness in horror film, and hipster cultures.

Ruth Hawthorn is Senior Lecturer in American Literature at the University of Lincoln. She is currently completing a monograph on American detective fiction for the BAAS Paperbacks series with Edinburgh University Press. Her research interests include crime fiction, the literature of LA and ecocriticism.

Nicole Kenley is a lecturer in the English Department at Baylor University, working primarily on the relationship between crime fiction and globalisation. She has published journal articles on detective fiction in *The Canadian Review of Comparative Literature, Mississippi Quarterly,* and *Clues.* Her work also appears in the edited collections *Crime Uncovered: Antihero*, *Teaching Crime Fiction*, and the *Routledge Companion to Crime Fiction,* as well as in the forthcoming *Cambridge Companion to World Crime Fiction.* She is also the co-editor, with Malcah Effron, of *The Journal of Popular Culture*'s forthcoming special issue on place, space and the detective narrative, as well as a reviewer for *Crime Fiction Studies.*

Michael Parrish Lee is Lecturer in English Literature and Creative Writing at Leeds Beckett University. He is the author of *The Food Plot in* *the Nineteenth-Century British Novel* (2016). His essays have appeared in *Novel: A Forum on Fiction*, *Nineteenth-Century Literature*, *Victorian Literature and Culture*, *Studies in the Novel*, and *Literary Theory: An Anthology* (2017), and his fiction has appeared in *Conjunctions*.

John Miller is Senior Lecturer in Nineteenth-Century Literature at the University of Sheffield, President of ASLE-UKI (Association for the Study of Literature and the Environment, UK and Ireland) and co-editor of Palgrave Studies in Animals in Literature. His books include *Empire and the Animal Body* (Anthem, 2014) and *The Heart of the Forest* (British Library Publishing, 2022).

Jordan Sheridan is an adjunct professor in the Animal Studies Department, New York University. His work focuses on representations of genetically altered animals in literature and science fiction.

Adrian Tait is an independent scholar and environmental critic with a particular focus on the nineteenth century. A long-standing member of the Association for the Study of Literature and the Environment (ASLE-UKI), he has regularly published in its journal, *Green Letters*. He has contributed to a number of other scholarly journals, including the *European Journal of English Studies* (2018), and to essay collections such as *Nineteenth-Century Transatlantic Literary Ecologies* (2017), *Victorian Ecocriticism* (2017), *Enchanted, Stereotyped, Civilized: Garden Narratives in Literature, Art and Film* (2018), *Perspectives on Ecocriticism* (2019), and *Literature and Meat Since 1900* (2019).

Sally West is Senior Lecturer in English Literature at the University of Chester. She teaches literature of the long nineteenth century, Shakespeare and crime fiction. Her research background is in Romantic literature, and she has published on Coleridge, Shelley and literary influence (*Coleridge and Shelley: Textual Engagement*, Ashgate, 2007). More recently, her research interests have moved to Crime Fiction. She is currently working on a research project on the significance of space and place in the work of Patricia Highsmith.

Giles Whiteley is Professor of English Literature at Stockholm University. He is the author of four monographs on topics ranging across nineteenth-century literature, most recently *The Aesthetics of Space in Nineteenth-Century British Literature, 1843-1907* (Edinburgh University Press, 2020).

Origins and Evolutions: The Brutal History of Detective Fiction

Ruth Hawthorn and John Miller

In the literary history of detective fiction, Edgar Allan Poe's "The Murders in the Rue Morgue" (1841) occupies a notable position. As the first of Poe's three Dupin mysteries, the story is in many accounts coeval with the emergence of the detective genre. According to Stephen Rachman, Poe gave "the form its initial shape" and "created its first great detective".[1] Consequently, as John Scaggs notes, a "critical consensus" has emerged in which "the detective story begins" with Poe, who, by this token, assumes the status of "the 'father' of the detective genre".[2]

If Poe is "the first truly modern exponent" of detective fiction, as Rachman phrases it,[3] the genre's first perpetrator—its ur-criminal—is the "the large fulvous Ourang-Outang of the East Indian Islands" that Dupin deduces is responsible for the murder of Madame L'Espanaye and her

R. Hawthorn (✉)
University of Lincoln, Lincoln, UK
e-mail: rhawthorn@lincoln.ac.uk

J. Miller
University of Sheffield, Sheffield, UK
e-mail: John.Miller@shef.ac.uk

R. Hawthorn, J. Miller (eds.), *Animals in Detective Fiction*, Palgrave Studies in Animals and Literature,
https://doi.org/10.1007/978-3-031-09241-1_1

daughter in the locked room in Paris at the mystery's heart.[4] In detective fiction's ostensibly inaugural text, at the crucial moment of Dupin's revelation of the murderer, the narrative resolution is assembled from the body of an animal. Here is how Dupin pieces together the brutal circumstances of the two women's murders. As the detective reconstructs events, he elicits two key insights from the unnamed narrator. Presented with a "little tuft" Dupin has disentangled "from the rigidly clutched fingers" of the deceased Madame L'Espanaye, the narrator is shocked by a sudden realisation: "'Dupin!' I said, completely unnerved; 'this hair is most unusual—this is no *human* hair'". Then, confronted by the impressions of fingers in the bruises on the victim's body, the narrator is led to a further discovery: "This … is the mark of no human hand" (260). The narrative pivots on the unexpected realisation of animal agency which Poe tellingly introduces in the negative. Who but a human could act with such will and force? What is startling to the narrator is not so much that the killer is an animal, but that it is not a human. In the task of assembling the facts of the case, this amounts to the same thing, but the phrasing (the repetition of "no human") is significant. Animality emerges as the absence of humanity; the animal functions as the shadow of the human in a way that foregrounds an underlying anthropocentrism at the story's decisive moment, even while (as we shall see) the narrative also does much to question and complicate conventional understandings of human–animal difference.

Since Poe, detective fiction has become a vast cultural enterprise, involving distinct styles, periodicities and regional particularities while leaking into a variety of other genres, the gothic most obviously, and informing the broader though closely related category of crime fiction in which the narrative focus is directed towards some kind of criminality, but without necessarily including the specific conventions of detection. To follow the language of Poe's narrator, detective fiction is no human genre; the moment of the genre's apparent inauguration in "The Murders in the Rue Morgue" announces a form of writing that emerges out of the imagining of human–animal relations and which, throughout its subsequent development, consistently stages and investigates conceptions of species. Detective fiction's evolution after Poe involves a huge array of animal presences in which creatures take many roles. The figure of the beastly criminal perpetrator endures beyond Poe both in the literal form of the animal as killer and in the figurative form of the (human) killer as animal: a degraded being deprived of the tenets of order and civilisation that are often expressed as synonymous with the idea of humanity. At the same time, the trope of the human as animal does not just concern the depiction of criminals. From the "sleuth-hound" Sherlock Holmes to Dashiell Hammett's Sam Spade with his tendency to grin "wolfishly", detection involves the development of beastly characteristics.[5] Detective fiction accommodates gentler and more law-abiding creatures too. Wilkie Collins, a foundational figure in the history of the British detective novel, depicts the villain Count Fosco in *The Woman in White* (1859) surrounded by his "pretties", "a cockatoo, two canary-birds and a whole family of white mice",[6] while Koko and Yum Yum, the feline sidekicks of Lillian Jackson Braun's popular *The Cat Who…* series from the 1960s, show animals living on the right side of the law. There are animal victims, like the "half-charred corpse of a cat" used to menace Blomkvist and Salander in Stieg Larsson's *The Girl with the Dragon Tattoo*,[7] and plenty of animal clues, such as "the curious incident of the dog in the night time" in Arthur Conan Doyle's "Silver Blaze" (a mystery about a missing race horse) in which a dog's failure to bark leads Holmes to the solution.[8] Detective fiction is a genre with an intimate attachment to narrative patterns and character typologies (the murder and its investigation; the detective, the witness, the perpetrator and the victim); animals and animalities exist across these structures.

A complete summary of the manifold ways in which animals feature in detective fiction would be a vast undertaking, though it is significantly easier to provide a summary of the critical field that has emerged in relation to this topic. Given how strongly detective fiction and animals connect, it is extraordinary how little concerted critical attention has been paid to the genre's non-human creatures. By now, this kind of observation has become a routine gesture for scholars concerned with the literary representation of animals. Animals are everywhere in literature of all kinds, but up until around the Millennium very little attention was paid to the ways in which understandings of literature might be enriched by analysis of its creatures, or, conversely, by the way in which human–animal relations might be more deeply understood by taking seriously questions around textual representation. The rise of literary animal studies provides for analysis of the multifarious animal presences that have been hitherto obscured by the dominant anthropocentric model of literary scholarship. Works like Poe's "The Murders in the Rue Morgue" have now attracted a good deal of reflection on the question of the animal; indeed, Poe's work could be considered as a seminal part of an emerging canon of literary animal studies. There have also been some relatively isolated accounts of the wider question of animals in detective texts in key works of animal studies, perhaps most notably Susan McHugh's discussion of service dogs and detective fictions in *Animal Stories.*[10] But the study of detective fiction has not formalised this agenda as part of its critical terrain, even if discussions of detective fiction and histories of science contain some oblique references to animals.[11]

Take, for example, the many critical companions to detective and/or crime fiction that have been published since the animal turn began to take shape. *The Cambridge Companion to Crime Fiction*—one of the more commodious books in that series of "authoritative guides" to key topics[12]—contains hardly a beast in the whole thing, even if the 2003 publication date places it right at the start of the period during which the study of animals in literature started to become part of institutional academic culture. More recently, *The Cambridge Companion to American Crime Fiction* (2010) is similarly devoid of more-than-human creatures. The same year saw the immense Wiley-Blackwell *A Companion to Crime Fiction* which despite no less than forty-seven chapters evades the question of the animal entirely. Even a book from 2015 promising *New Perspectives on Detective Fiction* remains steadfastly human in its range of priorities. A chapter on Agatha Christie's *The Mousetrap* and Tom Stoppard's spoof *The Real Inspector Hound* offers an intriguing hint of the animal that the argument leaves undeveloped. If the question of animals in detective fiction remains underexplored, the cognate field of ecocriticism has started to engage with the genre's involvement with ecological questions. The 2020 *Routledge Companion to Crime Fiction*, for example (at forty-five chapters, another behemoth), does include a chapter on "Crime Fiction and the Environment". More sustained engagement with ecological questions has appeared in a special edition of the journal *Green Letters* on "Crime Fiction and Ecology" which explores how "generic features offer opportunities to reflect on the forms and functions of environmental criticism and ecological narratives",[13] though there is little attention to the representation of animals throughout this collection.

The role of animals in detective fiction, then, is both critically marginal and historically integral. The aim of this volume is to begin to redress the curious lack of creatures in critical engagements with the detective form and in doing so to reconceive of the genre as one in which the question of species is key. This task carries both ethical and political weight. Animal-centred literary histories are concerned with foregrounding the multispecies dimensions of human cultural production. At all points human lives are involved with those of other species; literary studies can be a way to attend to the significance of other lives in a way that insists on the necessity of our ethical obligations towards them and aspires to a politics beyond human exceptionalism. Detective fiction can claim a particular significance in this enterprise. Literary animal studies has necessarily and unsurprisingly involved considerable attention to the depiction of violence against animals. So much of human social, economic and cultural activity involves animal death, either directly or indirectly; an early collection of essays, *Killing Animals* by the Animal Studies Group emerges out of the contention that "killing represents by far the most common form of human interaction with animals".[14] It is in this context that the analysis of detective fiction is particularly valuable for understandings of the literary representation of animals. Species violence is both a recurrent topic for detective fiction and an aspect of its narrative structure. It is certainly not the case that detective fictions are always—or even often—on the side of the animals, but analysis of the forms of violence the genre entails (and the wider imagination of species it is embroiled in) contributes to the task of unravelling anthropocentric logics and bringing the challenge of how to live well with other creatures into clearer focus. In what follows in this opening chapter, we turn again to Poe's "The Murders in the Rue Morgue", and then to its precursors, in order to show how the origins of detective fiction embed animals into the genre's formal structures and emerging cultural politics. We move next to another case study, Agatha Christie's *A Pocket Full of Rye* (1953) to develop a key point from the critical heritage around Poe's story: the splicing together of figurative and literal depictions of animals in detective fiction. In the final section, we outline how the fourteen essays of *Animals in Detective Fiction* move towards an emerging multispecies understanding of the genre.

Origins

"The Murders in the Rue Morgue" provides a multifaceted depiction of the ambiguities and anxieties surrounding the deceptively straightforward question of the difference between animals and humans. The fate of the ape unfolds through a narrative structure that disturbs conceptions of human–animal difference only to restore them at the tale's close. If the orang starts its misadventures acting like a human ("Razor in hand, and fully lathered … sitting before a looking-glass", 264), the story's ending puts it right back in its place. After the ape's responsibility is established, it receives no trial or formal sentencing; in accordance with a Cartesian philosophical tradition, the modern judicial process does not recognise animals as capable of the kind of agency that would constitute guilt. As Christopher Peterson summarises, legal accountability "presupposes … a consciousness that humans have historically denied to animals".[15] Instead, once the ape has been recaptured, it is sold on "for a very large sum" to the *Jardin des Plantes*. By these means it can be re-established in its accustomed roles as commodity, specimen and spectacle. The order of species hierarchy is restored. If the ape's transfer to the *Jardin des Plantes* might give the appearance of mercy (the ape at least escapes the death penalty a human would be destined for in these circumstances), any sense of clemency is fraught with irony. An ape on the run in Paris would be imprisoned anyway regardless of anything it may or may not have done; it is the fate of the animal in this context that it is always already available for punishment.

The familiar recourse to human exceptionalism is uneasily upheld in the story, however. After the ape has been carted off to its inevitable imprisonment, the story's final paragraph provides an intriguing parting animal image, albeit one that is easily missed as a seemingly throwaway remark with little bearing on the central narrative interest. Reflecting on the resolution of the case, Dupin has some unkind words to say about the disgruntled police Prefect, who in his lack of insight and application appears to be "all head and shoulders, like a codfish" (266). The simile is unusual, but the structure is familiar: it is in its failures or foibles that the human comes to resemble the animal ("stubborn as a mule" or "devious as a fox"). In the story's logic there is a world of difference between the playful banter about the Prefect (who remains after all, as Dupin goes on to say, "a good creature") and the ape's murderous spree, but there is a key point of similarity between the figurative fish and the material ape. Both express the idea of the human as structured through an ongoing negotiation with the idea of the animal in a way that complicates the story's anthropocentric narrative trajectory. We will return to the codfish momentarily, but first the ape.

The story's twist is premised on the assumption that the killer must be a human (hence the narrator's astonishment). Famously, the "shrill" or "harsh" voice heard by several people coming from the crime scene is identified by different witnesses as a speaker of various different nationalities, but by no one as an animal. "The fiendish jabberings of the brute" (90), as Dupin finally describes them, are variously understood as the speech of a Spaniard, Frenchman, German, Englishman, Russian or Italian.

By this misrecognition, the story places discourses of nationality and discourses of species in close proximity to one another. That the sound of an ape can be construed as the generic sound of the foreigner illustrates the way in which categories of otherness within the human operate with reference to questions of the animal. To hear an ape as a foreigner is to understand national difference through the lens of animality. Dupin appears to take the more charged question of race out of the equation: "You will say that it might have been the voice of an Asiatic—of an African", he comments, but goes on to explain that "Neither Asiatics nor Africans abound in Paris" (78). Crucially, this rather awkward disavowal has provided criticism with a strong invitation to read the ape in precisely the terms Dupin negates.

Although the ape of Poe's story is unequivocally an animal, there has long been a critical reluctance to read the orang-utan as *just* an animal. Rather, the ape has often been read as the figuring of the human *as* animal as a commentary on and embodiment of racial/racist anxiety in the wake of a series of riots in Philadelphia in the 1830s and 1840s where Poe was resident while composing the story. In short, as Ed White puts it, the story should be read as a "response to American slave rebellions".[16] The peril of the beast is understood in nineteenth-century America, by this token, to function as the peril of the slave freed from the restraints of white ownership, a trope that automatically understands race in relation to species in order to delegitimize movements for racial equality. Behind this pattern is a wider linkage between race, criminality and animality which finds its most influential expression in the conclusions of the Italian anthropologist Cesare Lombroso later in the nineteenth century. Crime, for Lombroso, is a "return to the early brutal egotism natural to primitive races".[17] In becoming criminal and leaving behind the strictures of law and ethics, the human is routinely thought to abandon part of its humanity and to become more akin to animals. Since the period's racist imaginary automatically understands black Americans as primitive and hence as bestial, criminality is integral to racial discourses. Taken as an allegory for anxieties about slavery, "The Murders in the Rue Morgue" constitutes a case study of the complex intersections of racial and species politics that encourages a critical examination of who exactly is designated by the Anthropos of anthropocentrism.

Importantly, this is not to say that the story is *only* about racial politics so that the murderous ape must be understood, as Peterson comments on White's argument, "as something other than an orang".[18] It is not just slave rebellions that comprise the immediate context of Poe's composition of the story; there were also "several popular primate displays in Philadelphia" during Poe's residence there which link the fictional orang to the material history of apes.[19] Rather than seeing slave rebellions and primate displays as opposing contexts supporting divergent readings, their common involvement with colonial history encourages an approach that sees them as complementary historical symptoms of a far-reaching hierarchical regime. Poe's ape is both allegory and materiality, a textual device and a depiction of a body in history. Indeed, Poe's depiction of the orang-utan would go on to have a notable influence of the cultural construction of the great apes.

When the French-American explorer Paul du Chaillu became the first Westerner to publish an account of a meeting with a gorilla in West-Central Africa in 1861, his narrative owed much to Poe. Where Poe has his orang "Gnashing its teeth, and flashing fire from its eyes" (89), du Chaillu notes how his gorilla's "eyes began to flash fiercer fire" when it stood before his gun.[20] Du Chaillu's work would do much to solidify the image of the devilish primate, a trope that endured to 1933's *King Kong* and beyond and which remains consistently tied to conceptions of race and empire. It becomes impossible to separate the allegorical reading of the ape as figure from the literal reading of the ape as animal; as Colleen Glenney Boggs has concluded, Poe's representation of the ape is "not only a symbolic act, but also a corporeal process".[21] Poe's orang is Derrida's *animot*, a complex hybrid of materiality and textuality. While the story's conclusion consigns the ape back to its structural location in the chain of being, it does so only after a fraught process of recognition and misrecognition has rendered the human as a site of ambivalence. The curious parting image of the codfish in this context intrudes an uncomfortable animal image back into the text at the moment of humanity's apparent triumph over the animal. Even when the mystery is solved and the ape is safely returned to its cage, the spectre of the animal continues to haunt the human.

Poe's complication of the idea of the human in "The Murders in the Rue Morgue" gains emphasis from a consideration of the prehistory of the detective form. If Poe's pre-eminence in histories of detective fiction represents a critical orthodoxy, it is also a conventional critical gesture to offer a series of caveats that explain how the genre's most recognisable characteristics exist in prototype in earlier texts. The most obvious caveat is Poe's debt to William Godwin, whose 1793 novel of tyranny and injustice *Caleb*

Williams provided the later writer with a model of narrative method. Poe asserted that it is through adopting Godwin's plan of working "with the dénouement constantly in view that we can give a plot its indispensable air of consequence" (although Godwin's composition of the novel in reality was nothing like so well thought out).[22] Certainly, as Michael Cohen summarises, *Caleb Williams* "has the crime story's distinctive construction" [23] with an emphasis, in Julian Symons' terms, on "murder, its detection, and the unrelenting pursuit by the murderer of the person who has discovered his guilt".[24] Outside the Anglophone tradition, Voltaire's *Zadig* (1747), with its depiction of the eponymous hero's "peculiar Talent to render Truth as obvious as possible" through careful observation, is frequently considered to underlie the ingenious deductions of Poe's detective C. Auguste Dupin.[25] Dorothy L. Sayers, an important historian of detective fiction as well as a contributor to the genre, located the early seeds of the detective form further back in "Oriental folk-tales, in the Apocryphal Books of the Old Testament, in the Play-Scene in *Hamlet*" and in certain passages of Aristotle's *Poetics*.[26] More recent attention in Anglophone criticism to literary histories beyond Europe and America has solidified Sayers' generalising reference to "Oriental folk-tales". What Sayers has in mind here is the Middle Eastern collection *One Thousand and One Nights*, which dates back at least to the ninth century, and anticipates the narrative structure of detective fiction in "The Tale of the Three Apples", for instance. Beyond Sayers' frame of reference, it is worth noting the development of detective fiction in China, most significantly in Gong'an fiction, usually concerning the adventures of magistrates, which began to take shape as early as the tenth century. The origin of detective fiction is perhaps best conceived, therefore, as historically and geographically dispersed in a way that complicates the attribution of the genre's origin to a single author. Just as the established literary history of detective fiction turns to Poe's precursors to discover the networks of influence behind the genre's formal properties, so an animal-centred literary history can move back through earlier texts to discern how the animal tracks in "The Murders in the Rue Morgue" are already part of the traditions Poe draws on (and indeed part of the traditions he is not able to draw on). The beastly origins of detective fiction run deep.

Voltaire's illustration of the detective's ratiocinative faculties in *Zadig* for example demonstrates the significance of animals in the formulation of detective method. Having retired "to a little Country House on the Banks of the *Euphrates*", the eponymous hero whiles away his time examining "the Nature and Properties of Animals and Plants" so that he becomes "capable of discerning a Thousand Variations in visible Objects, that others, less curious, imagin'd were all alike".[27] Consequently, when the Queen's spaniel goes missing, Zadig is able to exactly reconstruct the dog's appearance without having encountered her, deducing, merely from the pattern of the traces in the sand, that she has recently had puppies, had long ears and is lame in "her left Fore-foot". Then, confronted with the appearance of a series of "Prints made upon the Sand by a Horse's Shoes", Zadig is able to reconstruct the exact proportions of the King's missing palfrey with the kind of miraculous facility for deduction that would become a hallmark of Sherlock Holmes, most famously.

T. H. Huxley, among the nineteenth century's most prominent thinkers about questions of the animal and the human, found in Voltaire's novella a striking example of natural historical method, suggesting that:

> In no very distant future, the method of Zadig, applied to a greater body of facts than the present generation is fortunate enough to handle, will enable the biologist to reconstruct the scheme of life from its beginning, and to speak as confidently of the character of long extinct beings, no trace of which has been preserved, as Zadig did of the queen's spaniel and the king's horse.[28]

The link between detective fiction and natural science is a well-remarked aspect of the genre's literary history with important implications for reading the genre's non-human creatures. The natural historical context provides a direct link from Voltaire to Poe via the French zoologist Georges Cuvier. Cuvier cites *Zadig*, as Huxley observes, in "one of the most important chapters of [his] greatest work";[29] thus, as Lawrence Frank summarises, "Cuvier placed himself within the empirical tradition that Voltaire promoted in *Zadig*".[30] Poe then has Dupin lead the narrator towards the mystery's solution via "a passage from Cuvier" which contains "a minute anatomical and generally descriptive account of the large fulvous Ourang-Outang of the East Indian Islands". Dupin's skill in "The Murders in the Rue Morgue" resides in his ability to recognise a beast, where others had recognised man. As such, it is scientific method that establishes order out of chaos and in doing so reasserts the priority of the human. The significance of animals in detective writing is not just a matter of what comes to be known—the crimes the ape is found to have committed—but also a matter of how knowledge is constructed and legitimated: the epistemological methods by which the human knows the animal and comes to assure itself of its own pre-eminence. The very form of the detective story encodes the interplay between human rationality and animal instinct or appetite in a way that invites analysis in terms of the cultural politics of species.

Moving forward through Poe's progenitors, animals in *Caleb Williams* form a crucial part of the novel's depiction of its cultural context, narrative drama and, most importantly for a political writer like Godwin, exploration of social injustice. As a novel concerned primarily with life "in a remote county of England",[31] animals are everywhere, even if their appearance can seem incidental, as in the consistent reference to horses as the necessary mode of transport, or in the "arts of shooting, fishing, and hunting" (19) that are a routine element of country life. Yet these practices are not unconnected to the novel's primary political agenda. When the merciless landowner Tyrrel prosecutes his tenants the Hawkins family, the trumped-up charge he brings is based on the notorious 1723 Black Act, through which a number of crimes against property came to be capital offences. Godwin quotes a clause from the law that revolves around the proper form for human–animal relations:

> any person, armed with a sword, or other offensive weapon, and having his face blackened, or being otherwise disguised, appearing in any warren or place where hares or conies have been or shall be usually kept, and being thereof duly convicted, shall be adjudged guilty of felony, and shall suffer death (78).

Relationships with animals are an important way in which social structures are performed and policed in the text. Moreover, conventional rural pursuits function in the novel's figuring of its own narrative structure in a way that uses human–animal relationships to bring the novel's form and politics into alignment. As Williams evades his hunters, he compares his experience of persecution to the "terrors of an animal that skulks from its pursuers" (265). In being refused the possibility of justice, Williams is refused his humanity and forced to exist beast-like on the fringes of society, an experience that emerges forcefully in the novel's original, unpublished ending. Languishing in prison and driven to madness, Williams concludes his narrative with the resonant assertion, "HERE LIES WHAT WAS ONCE A MAN!" (346) The ultimate destination of the intrigue of *Caleb Williams* is to empty the protagonist of his humanity. For Godwin, the language of species is key to the novel's ethical and ideological commitments and central to the trope of pursuit that constitutes much of the novel's value for the development of detective fiction. Detection would remain throughout its subsequent evolution consistently coded through the rhetoric of the hunter and the hunted and predator and prey. The common detective ruse of the red herring is a prominent way in which this history remained coded into the formal structures of the genre; William Cobbett is thought to have coined the term in 1807 with reference to his use of a fish to distract a pack of hounds from a hunt for a hare.[32]

Moving beyond Poe's European progenitors to the *The Thousand and One Nights*, animals play a subtle but structurally important role in "The Story of the Three Apples". The story focuses on a poor fisherman who is struggling to feed his family. His luck appears to have turned when he casts his net and dredges up a heavy, locked case which he sells for "a hundred pieces of gold".[33] When the chest is opened in the palace it is found to contain the remains of a murdered woman. While the subsequent investigation involves little engagement with non-human creatures, the story's premise is closely involved with questions concerning animals: It is in the moment of human–animal contact (the fisherman casting his net) that the intrigue is initiated. To turn briefly to China, the eighteenth-century *Celebrated Cases of Judge Dee*, translated into English by the Dutch sinologist and ape collector Robert van Gulik, like "The Murders in the Rue Morgue" involves the discovery of an animal perpetrator. In "The Case of the Poisoned Bride" a young woman dies suddenly on her wedding night, killed it would appear by poison in her tea. It turns out that as a maid prepared the bridal feast an "adder nestling in the mouldering beam was attracted by the hot steam and its venom dropped in the pan underneath".[34] For readers most familiar with a European tradition of detective writing, the snake as killer (albeit here unintentionally) anticipates Arthur Conan Doyle's 1892 story "The Adventure of the Speckled Band" in which Holmes deduces that murder has been committed through the agency of "a loathsome serpent" that has been deployed as a weapon by Grimesby Roylott.[35]

This brief sketch of the origins of detective fiction from Poe's inauguration of the genre back through its prehistory enables three key conclusions. Firstly, the genre's gravitation towards animals should not be read as merely incidental; it would be difficult after all to identify any genre of writing that was so resolutely human it contained no trace of other creatures. Rather, it constitutes a central part of the conception of the genre's formal properties, through the trope of the detective as hunter, the elision of criminality and animality, and the form's debt to the humanist tradition of natural science. Secondly, the genre's politics—its involvement with questions of race, class, gender and more—operates in relation to its engagement with animals. Lastly, a globalised focus brings out other conceptions and histories of animal signification which intersect with a Euro-American tradition, but which emerge from a historically distinct lineage.

"The Murders in the Rue Morgue" encourages a multispecies approach to the history of detective fiction in a very explicit way; there is no avoiding the ape at the heart of the narrative. The beastly origins and prehistory of the genre also encourage an animal-centred approach to texts in which non-human creatures appear to be marginal. Agatha Christie is perhaps the best-loved and most widely read example of a socially conservative strand in the history of detective fiction. As Scaggs summarises, "Christie's consumers … consisted of respectable, suburban readers …who shared Christie's upper middle-class, property-owning, bourgeois ideology, and were keen to have it confirmed".[36] An animal-focused approach to her 1953 novel, *A Pocket Full of Rye* reveals how this ideology revolves not just around violence against animals, but also becomes embroiled in the prospect of global ecological catastrophe.

Evolutions

Although animals do not supply the central narrative drama in *A Pocket Full of Rye*, they are nonetheless an important and complex clue in the unravelling of the mystery. The novel begins with the sudden death of the businessman Rex Fortescue, poisoned by taxine—a toxic alkaloid derived from the yew tree—which had been added to his breakfast marmalade. At the head of the investigation is Inspector Neele who travels to the Fortescues' country home Yewtree Lodge to track down the killer. Before too long, a second murder adds to Neele's task: Rex's young trophy wife Adele succumbs to cynanide in her tea. Shortly afterwards, the maid Gladys is found strangled in the grounds. Joined by the iconic detective figure of Miss Marple, Neele begins to suspect a connection between the murders and some historic bad blood between Fortescue and the McKenzie family: Fortescue had apparently tricked his partner McKenzie out of a mining concession in East Africa in his youth and left him there to die. Although the McKenzies are part of the plot—it turns out that Jennifer Fortescue, the wife of Rex's son Pervical, is McKenzie's vengeful daughter—they are eventually established as a decoy. The murderer is revealed by Miss Marple as Fortescue's other son, the disinherited Lance who, newly married, returns from Africa to Britain in search of riches. Lance seduced Gladys into poisoning Rex, under pretences of becoming her husband (and later murdered her to keep her quiet), before killing Adele to stop her inheriting Rex's wealth.

The function of animals in the narrative derives from the nursery rhyme which appears to hold the key to the mystery. A curious facet of Rex Fortescue's demise is that at his death his suit pocket contains a handful of rye. Adele met her end while "eating bread and honey" (92) while Gladys "*was in the garden hanging out the clothes*" (91, original emphasis). That Rex means King, and Gladys had a peg clipped onto her nose after her death confirms that the murderer is acting out the sequence of the second stanza of "Sing a Song of Sixpence":

> The king was in his counting house, counting out his money;
> The queen was in the parlour, eating bread and honey.
> The maid was in the garden, hanging out the clothes,
> When there came a little dickey bird and nipped of her nose (91).

What this means, as Miss Marple explains, is that "there must *be* blackbirds" in the case (92, original emphasis) in order to fulfil the rhyme's first stanza:

> Sing a song of sixpence, a pocket full of rye.
> Four and twenty blackbirds, baked in a pie.
> When the pie was opened the birds began to sing;
> Wasn't that a dainty dish to set before the king? (91)

Sure enough Neele and Miss Marple uncover plenty of intriguing blackbirds.

Two events involving blackbirds are found to have occurred in the lead up to Rex Fortescue's death. Firstly, there is what the housekeeper Mary Dove describes as "that silly business" (99):

> Four dead blackbirds were on Mr Fortescue's desk in his study here. It was summer and the windows were open, and we rather thought it must have been the gardener's boy, though he insisted he'd never done anything of the kind. But they were actually blackbirds the gardener had shot which had been hanging up on the fruit bushes (99).

A second incident makes more specific reference to "Sing a Song of Sixpence" as Neele's colleague Sergeant Hay declares he has "found out about the blackbirds":

> in a pie they were. Cold pie was left out for Sunday night's supper. Somebody got at that pie in the larder or somewhere. They'd taken off the veal and 'am what was inside it, and what d'you think they put in instead? Some stinkin' blackbirds they got out of the gardener's shed (120).

Together these macabre events constitute the central material presence of animals in Christie's novel. Although Marple and Neele concern themselves greatly with the meaning of the blackbirds, they are not, it turns out, directly related to the murder. It was not the killer Lance who was responsible for these cruel jokes, but rather Jennifer who, in seeking to revenge her family against Rex Fortescue for the death of her father, hoped merely to give Fortescue "a kind of fright" (182). The significance of the blackbirds is based on a linguistic coincidence. The mine in East Africa that created the fatal breach between McKenzie and Fortescue was known as the Blackbird Mine; Jennifer's prank is designed as an ominous reminder of Fortescue's dubious exploits.

The bodies of the blackbirds instead of forming part of the murderer's plot are part of his cover story. As Marple explains, "all this blackbird business is a complete *fake*. It was *used*, that was all, used by someone who heard about the blackbirds—the ones in the library and the ones in the pie" (176, original emphasis). On one level, the significance of "Sing a Song of Sixpence" is illusory. On another level, the blackbird—or at least the word Blackbird—remains crucial to the killer's motivation and to Marple and Neele's solution of the mystery. The Blackbird Mine had initially been taken on by Rex Fortescue as a gold mine, but the site had proved to be worthless. McKenzie's death occurred in pursuit of an empty dream. Years later the mine was discovered to contain uranium. Lance realised that there was "a fortune to be grasped" (180) and so his murderous plans were born. Here Christie touches, albeit fleetingly, on an important historical context. The 1950s saw a significant increase in global demand for uranium as the cold war nuclear arms race took hold. The Blackbird Mine's location in Tanganyika (today Tanzania) places it at the site of significant uranium deposits which attracted occasional media coverage in the years preceding the publication of *A Pocket Full or Rye*. There is a notable historical irony behind the novel's characteristically cosy conclusion. As Miss Marple returns home at the start of the final chapter, her maid Kitty notes comfortingly "you'll find everything very nice in the house. Regular spring cleaning I've had" (186). Christie's insistence on bourgeois social order is eerily juxtaposed by the weapons of mass destruction that provide the ultimate reference point behind the nursery rhyme's "Four and twenty blackbirds".

That *A Pocket Full of Rye* provides multiple ways of signifying blackbirds is not lost on Neele when he asks Jennifer for any information she might have about blackbirds "Alive or dead or even, shall we say, symbolical" (109). Neele's remark tellingly encapsulates a series of shifts in the signification of the birds throughout the novel. In the first instance, although the text affords this no space, the living, material birds are involved in pecking at the fruit bushes, which is presumably the reason why the ground staff kill them. Once dead the birds are deployed by Jennifer as a signifier of her family's tragic history in East Africa. It is this "symbolical" function that Lance seizes on to construct the elaborate murder plot whereby the use of the birds to refer to the mine in Tanganyika is made to refer to the nursery rhyme. Although the image of the blackbirds baked into a pie is based on the historical practice of the *entremet*—a form of culinary entertainment that involved springing living blackbirds from pastries among many other devices—the novel presents the rhyme as a riddle in which the connection to material birds is increasingly distant. The birds are made to gesture away from themselves to the novel's central human narrative. That, after all, is the function of the clue: to allow other narratives to form. If, as John Scaggs notes, one of the defining characteristics of detective fiction is that it "foregrounds … reading as a quest for meaning",[37] the effect of this in *A Pocket Full of Rye* is to contain the blackbirds' meaning in a network of textual references that is premised on killing. What this creates is three levels of violence: the killing of the birds, the epistemic violence of their conscription into other narratives as sign not body, and then, unspokenly but compellingly the overarching historical violence that the Blackbird Mine signifies: the looming menace to all planetary life in the developing arms race.

In this context, the novel's final image provides a complex take on the signification of animals in the detective form. Looking back on the case, Miss Marple experiences a "surge of triumph" akin to that "some specialist might feel who has reconstructed an extinct animal from a fragment of jawbone and a couple of teeth" (188). Most straightforwardly, the image evokes the genre's long-standing connection to natural history: piecing together the mystery is analogous to the taxonomist's task of assembling the truth of the natural world. More interestingly, this is an act of recovery that is premised on death: the whole mystery is based on the corpses of the blackbird and, macrocosmically, the idea of detection is based on how a specialist might reconstruct an extinct animal. There is a striking tension between the intellectual tradition Christie evokes here and the image's ethical and affective energy. Animal death is woven into the imagining of the genre's history and shadows the future the novel leaves unspoken. That Marple's overriding emotion at this point is one of triumph illustrates a stark negation of any question of the meaning of animal lives and worlds.

As the genre develops there are some notable texts which repurpose the form to a pro-animal agenda. We might think here of Carl Hiaasen's ecological fictions or the Nobel Laureate Olga Tokarczuk's *Drive Your Plow Over the Bones of the Dead* (2009) in which the mystery is based on a woman's desire to protect animals in the Polish forest from the violent machismo of the local hunters. The task of this volume is to bring the consideration of animals as meaningful ethical and ecological beings more fully to the fore of a form in which dying and killing seem their most usual roles.

Animals in Detective Fiction

The essays in this volume address the representation of animals in detective fiction in Anglophone texts from the late nineteenth century to the present day. These chronological and linguistic restrictions are determined not by the habitations of textual creatures: as we have seen, there are plenty of animals in earlier versions of the detective form and in traditions beyond the Anglophone world. Rather, the focus aims to begin a wider critical conversation on animals and detective fiction through an initial engagement with the modern, Anglophone tradition; we hope that a fuller focus on animals in global detective fictions can develop out of the more limited purview of this current volume. The essays are organised around four overlapping themes which address key topics in animal studies.

Part I "Ontologies" addresses questions of human and animal being and how detective fiction engages with structures of difference and hierarchy across species lines. In "Tigers, Criminals, Rogues: Animality in Dickens' Detective Fiction", Giles Whiteley explores the Victorian stereotype of representing criminals as animals in *Bleak House* (1852–53) and

The Mystery of Edwin Drood (1870). Whiteley examines firstly the links that connect the tiger and the tiger-hunter in Dickens' imagination before exploring the way in which the depiction of birds works to "deconstruct the perceived boundaries between those who represent the law and those who transgress it". What the analysis reveals is not a series of rigid or reductive binaries between different categories of being, but a "chiastic structure of mutual co-implication". Following Whiteley's discussion of Dickens, Adrian Tait investigates "Quantum Entanglements in Arthur Conan Doyle's *The Hound of the Baskervilles*". Just as Dickens appears to offer a reductive oppositional structure between human and animal, Conan Doyle's classic novel seems to offer a dichotomy between the great detective and his canine antagonist. Drawing on the new materialism of Karen Barad and Donna Haraway, Tait "highlights the parallels and connections between Holmes and hound" to posit "a doubling grounded in their mutual exposure to the agential presence of Dartmoor itself". From Conan Doyle, we turn to Nathan Ashman's chapter "Wolverines, Werewolves and Demon Dogs: Animalism, Criminality and Classification in James Ellroy's 'L.A. Quartet'". Ellroy's writing frequently presents a close link between animality and deviancy, a pattern which is "forcefully energised by rigid racial, sexual and political classifications". Ashman demonstrates how the Quartet at the same time raises questions about the human/animal binary by revealing animals as "victims of monstrous human behaviours". Ultimately, what emerges in Ellroy's work "is a ruthless and detached post war industrial culture that transforms both animal and human life into useable commodities".

Part II on "Ethics" concerns the ways in which detective fiction involves representations of ethical questions around animals. Although the depiction of animals in this context often foregrounds violence—violence done both to and by animals—there remain notable points of complexity in the genre which straightforward notions of victimhood and aggression do not accommodate. The first chapter in this section is John Miller's "The Psittacine Witness: Parrot Talk and Animal Ethics in Earl Derr Biggers' *The Chinese Parrot* and Earl Stanley Gardner's *The Case of the Perjured Parrot*". The phenomenon of talking parrots has long been used to think through questions around human–animal difference; while human speech is taken to reveal the superiority of human cognitive function, parrot talk merely suggests the lower facility of imitation. Against this philosophical background, Miller examines how two American detective novels use the plot device of the parrot witness in a way that encourages not just reflections on age-old questions about human–animal difference, but also an engagement with questions of animal ethics that focus particularly on the representation of grief and the parrot's function as a global capital. We then move from parrots to bears in Karin Molander Danielsson's chapter "Ecology, Capability and Companion Species: Conflicting Ethics in Nevada Barr's *Blood Lure*". In this essay, Danielsson returns to the motif of the animal as murderer in a novel which focuses on the depredations of a grizzly bear in the United States' Glacier National Park. Danielsson explores how Barr's depiction of animals involves a complex interplay between three distinct ethical positions: Donaldson and Kymlicka's conception of ecological ethics, capability ethics as set out by Martha Nussbaum, and Donna Haraway's companion species ethics. The discussion of competing ethics continues in Jordan Sheridan's "Laboratory Tech-Noir: Genre, Narrative Form and the Literary Model Organism in Jay Hosking's *Three Years with the Rat*". In addressing an sf-inflected noir mystery that develops in the context of scientific experimentation, Sheridan aims to reconceptualise the relationship between scientists and the animals they work on by thinking through how the novel's narrative strategies require the characters to cooperate with the novel's rodent protagonist Buddy in a complex and reciprocal way. Ultimately, Sheridan argues, the novel challenges "narratives of model organisms" that see them reductively "as inert, passive scientific objects". The final chapter in this section is Nicole Kenley's "Reptiles, Buddhism, and Detection in John Burdett's *Bangkok 8*". Here, Kenley examines Burdett's portrayal of "the Buddhist detection methodology of his protagonist, Royal Thai Police Detective Sonchai Jitpleecheep", in the broader context of Western adaptations and appropriations of Buddhist practices. Exploring the cyclical concept of *samsara*, Kenley argues that bringing this (potentially) counter-anthropocentric perspective to bear on the practice of detection has the effect of extending ethical considerations of justice to include non-human animals, while also questioning the grounds of the distinction between humans and animals.

Part III "Politics" explores animals in the detective form in terms of the ways in which power and responsibility are structured along species lines. In Michael Parrish Lee's "Animals, Biopolitics, and Sensation Fiction: M. E. Braddon's *Lady Audley's Secret*", the focus is on the biopolitics of human–animal relations. Specifically, Lee is concerned with how the novel's detective figure is embroiled in a contradictory attitude to animals. On the one hand, he is established as someone who cares deeply for animal life; on the other hand, the novel refers to his predilection for meat. If this seems to suggest that animals remain disposable, Lee finds a more nuanced pattern in Braddon's novel in which the abandonment of animal life in favour of human interests is resisted. Next, Sally West engages with "'The Motto of the Mollusc': Patricia Highsmith and the Semiotics of Snails". Highsmith had a long-standing personal relationship with snails, keeping them as pets and even "taking them to cocktail parties and other events in an oversized handbag". In her novel *Deep Water* and two short stories from *Eleven,* Highsmith uses snails as a way to critique the logic of commodification in the context of American capitalism. Taking the phrase "motto of the mollusc" from Gaston Bachelard's *The Poetics of Space*, West examines how Highsmith uses snails to signal human alienation from their environments and to reverse "the established human/ animal power dynamic". The final chapter in the "Politics" section is "'Before the white man came, when animals still talked': Colonial Creatures in Sherman Alexie's *Indian Killer* and Adrian C. Louis's *Skins*" from Alexandra Hauke. Hauke is concerned with a specific Native American tradition of detective writing, infused by animal mythology and, in particular, by the figure of the trickster. *Indian Killer* and *Skins* counter the colonial politics that often besets representations of Indigenous peoples and animals and in doing so provides a differently constituted version of the detective form which "prioritizes self-discovery and self-liberation from racially-marked violence over detection as a singular plot element".

The book's final part "Forms" investigates the significance of animals in the formal evolutions of detective fiction, encompassing the genre's crossover into other traditions as well as its development in response to other textual innovations and an increasing self-consciousness of its own formal devices. In "Aping the Classics: Terry Pratchett's Satirical Animals and Detective Fiction", Briony Frost shows how Pratchett's fantasy draws on and disturbs aspects of classic and hardboiled detective writing. Particularly, Frost explores the reimagining of the detective as animal in the City Watch octet and the ways in which this adjusts the conventional "constructions of the human criminal as animal". In *Guards Guards!* it is not the figure of the criminal who represents the animal in human form; rather, animality exists within the agents of human law as a key element of the structure of justice. In Joseph Anderton's "Animal Image and Human Logos in Graphic Detective Fiction", we turn from fantasy to the graphic novel in a chapter that retains a focus on the detective as animal. Anderton's chapter concerns two comic book detective series with animal protagonists: the noir cat investigator in Juan Diaz Canales and Juanjo Guarnido's Blacksad trilogy, and the badger detective in Bryan Talbot's *Grandville*. While both of these texts tap into a dualistic understanding of "animals as body/image and humans as mind/text", Anderton shows how these texts' anthropomorphic revision of the detective genre disturbs the hierarchy between "animal image and human logos". The volume's final chapter is Ruth Hawthorn's "'As easy to spot as a kangaroo in a dinner jacket': Animetaphor in Raymond Chandler and Jonathan Lethem". The focus of this chapter is Lethem's appreciative sf-noir pastiche, *Gun, with Occasional Music*. Drawing on Akira Mizuta Lippit's work on animetaphor, Hawthorn explores Jonathan Lethem's literalising of Chandler animalistic similes, arguing that the later text highlights the conservative power dynamics of the original while offering a more radical vision of human/animal slippage.

Notes

1. Rachman, "Poe and the Origins of Detective Fiction", 17.
2. Scaggs, *Crime Fiction*, 7.
3. Rachman, "Poe and the Origins of Detective Fiction", 17.
4. Poe, "Rue Morgue", 261. Subsequent references in parenthesis.
5. Conan Doyle, "The Red-Headed League", 64; Hammett, *The Maltese Falcon*.
6. Collins, *The Woman in White*, 222.
7. Larsson, *The Girl with the Dragon Tattoo*, 357.
8. Conan Doyle, "Silver Blaze", 23.
9. McHugh, *Animal Stories*, 27–64.
10. See, for instance, Lawrence Frank's work on evolution in *Victorian Detective Fiction* and Christopher Pittard's work on vivisection in *Purity and Contamination*.
11. Cambridge Companions, July 19, 2021. https://www.cambridge.org/core/what-we-publish/collections/cambridge-companions.
12. Walton and Walton, "Crime Fiction and Ecology", 3.
13. Animal Studies Group, *Killing Animals*, back matter.
14. Peterson, "Aping Apes", 151.
15. White, "The Ourang-Outang Situation", 8.
16. Lombroso, *Primitive Man*, 119
17. Peterson, "Aping Apes", 158.
18. Peterson, "Aping Apes", 154.
19. Du Chaillu, *Equatorial Africa*, 100.
20. Boggs, *Animalia Americana*, 114.
21. Poe, "The Philosophy of Composition".

22. Cohen, "Godwin's *Caleb Williams*", 203.
23. Cited in Cohen, "Godwin's *Caleb Williams*", 203.
24. Voltaire, *Zadig*, https://www.gutenberg.org/files/18972/18972-h/18972-h.htm#Ch_6.
25. Sayers, "Introduction", vii.
26. Voltaire, *Zadig*, https://www.gutenberg.org/files/18972/18972-h/18972-h.htm#Ch_6.
27. Huxley, "On the Method of Zadig", https://www.gutenberg.org/files/2627/2627-h/2627-h.htm.
28. Huxley, "On the Method of Zadig", https://www.gutenberg.org/files/2627/2627-h/2627-h.htm.
29. Frank, Frank, Lawrence. *Victorian Detective Fiction* p. 23.
30. Godwin, *Caleb Williams*, 5. Subsequent references in parenthesis.
31. "Red Herring", July 19, 2021. https://politicaldictionary.com/words/red-herring/.
32. "The Story of Three Apples", *The Thousand and One Nights*, July 12 2021, https://www.gutenberg.org/files/34206/34206-h/34206-h.htm.
33. Anon, *Judge Dee*, 169.
34. Conan Doyle, "The Speckled Band", 195.
35. Scaggs, *Crime Fiction*, 38. For a counter-reading, arguing against the idea that Christie and her contemporaries were cosily conventional, see Martin Edwards, *Golden Age*.
36. Scaggs, *Crime Fiction*, 74.

Works Cited

Allan, Janice, Jesper Gulddal, Stewart King, and Andrew Pepper, eds. *The Routledge Companion to Crime Fiction*. Abingdon: Routledge, 2020.

Animal Studies Group. *Killing Animals*. Champaign, IL: University of Illinois Press, 2006.

Anon. "The Story of Three Apples". *One Thousand and One Nights*. Translated Edward William Lane, July 12 2021. https://www.gutenberg.org/files/34206/34206-h/34206-h.htm

Anon. *Celebrated Cases of Judge Dee: An Authentic Eighteenth-Century Chinese Detective Novel*. Translated Robert Van Gulik. New York: Dover, 1976.

Boggs, Colleen Glenney. *Animalia Americana: Animal Representations and Biopolitical Subjectivity*. New York: Columbia University Press, 2013.

"Cambridge Companions". Cambridge University Press, July 19, 2021. https://www.cambridge.org/core/what-we-publish/collections/cambridge-companions.

Christie, Agatha. *A Pocket Full of Rye*. London: Fontana, 1958.

Cohen, Michael. "Godwin's Caleb Williams: Showing the Strains in Detective Fiction." *Eighteenth-Century Fiction* 10, no. 2 (1998): 203–220.
Collins, Wilkie. *The Woman in White*. Oxford: World's Classics, 2008.
Conan Doyle, Sir Arthur. "The Red-Headed League". In *The Adventures of Sherlock Holmes*, ed. Richard Lancelyn Green, 49–74. Oxford: World's Classics, 1993a.
Conan Doyle, Sir Arthur. "The Adventure of the Speckled Band". In *The Adventures of Sherlock Holmes*, ed. Richard Lancelyn Green, 171–197. Oxford: World's Classics, 1993b.
Conan Doyle, Sir Arthur. "Silver Blaze". In *The Memoirs of Sherlock Holmes*, ed. Christopher Roden, 3–29. Oxford: World's Classics, 1993c.
Cothran, Casey, and Mercy Cannon, eds. *New Perspectives on Detective Fiction: Mystery Magnified*. Abingdon: Routledge, 2015.
Du Chaillu, Paul Belloni. *Explorations and Adventures in Equatorial Africa: With Accounts of the Manners and Customs of the People, and of the Chace of the Gorilla, the Crocodile, Leopard, Elephant, Hippopotamus and Other Animals*. London: John Murray, 1861.
Edwards, Martin. *The Golden Age of Detective Fiction*. London: HarperCollins, 2016
Frank, Lawrence. *Victorian Detective Fiction and the Nature of Evidence: The Scientific Investigations of Poe, Dickens, and Doyle*. London: Palgrave, 2003.
Godwin, William. *Caleb Williams*. London: Penguin Classics, 1988.
Hammett, Dashiell. *The Maltese Falcon*. London: Orion, 2005.
Huxley, T. H. "On the Method of Zadig". July 12 2021. https://www.gutenberg.org/files/2627/2627-h/2627-h.htm
Larsson, Stieg. *The Girl with the Dragon Tattoo*. London: Quercus, 2008.
Lombroso, Cesare. *Criminal Man, according to the Classification of Cesare Lombroso*. New York: Puttnam's, 1911.
McHugh, Susan. *Animal Stories: Narrating Across Species Lines*. Minneapolis: University of Minnesota Press, 2011.
Nickerson, Catherine Ross, ed. *The Cambridge Companion to American Crime Fiction*. Cambridge University Press, 2010.
Peterson, Christopher. "The Aping Apes of Poe and Wright: Race, Animality, and Mimicry in 'The Murders in the Rue Morgue' and *Native Son*." *New Literary History* 41, no. 1 (2010): 151–171.
Pittard, Christopher. *Purity and Contamination in Late Victorian Detective Fiction*. Abingdon: Routledge, 2011.
Poe, Edgar Allan. "The Murders in the Rue Morgue". In *The Selected Writings of Edgar Allan Poe*, ed. G. R Thompson, 239–266. New York: Norton, 2004.
Poe, Edgar Allan. "The Philosophy of Composition", July 19, 2021. https://www.bartleby.com/109/11.html

Rachman, Stephen. "Poe and the Origins of Detective Fiction". In *The Cambridge Companion to American Crime Fiction*, ed. Catherine Ross Nickerson, 17–28. Cambridge: Cambridge University Press, 2010.
"Red Herring". Political Dictionary, July 19, 2021. https://politicaldictionary.com/words/red-herring/.
Rzepka, Charles J., and Lee Horsley, eds. *A Companion to Crime Fiction*. Chichester: John Wiley & Sons, 2010.
Sayers, Dorothy L. "Introduction" in *Great Tales of Detection: Nineteen Stories*. London: Dent, 1936.
Scaggs, John. *Crime Fiction*. Abingdon: Routledge, 2005.
Tokarczuk, Olga. *Drive Your Plow Over the Bones of the Dead*, translated Antonia Lloyd Jones. London: Fitzcarraldo, 2009.
Voltaire. *Zadig/L'Ingénu*, 12 July 2021, https://www.gutenberg.org/files/18972/18972-h/18972-h.htm#Ch_6.
Walton, Jo Lindsay, and Samantha Walton. "Crime Fiction and Ecology". *Special Edition of Green Letters: Studies in Ecocriticism* 22 (2018), 1–6.
White, Ed. "The Ourang-Outang Situation". *College Literature* (2003): 88–108.

ONTOLOGIES

Tigers, Criminals, Rogues: Animality in Dickens' Detective Fiction

Giles Whiteley

In his early essay "On Truth and Lying in a Non-Moral Sense" (1873), Friedrich Nietzsche contends that an animal cannot be a criminal: the commission of a crime means an awareness of breaking the law, a product of human culture. Connectedly, Nietzsche holds that human beings are distinguished from animals by their penchant for dissimulation. Lying is

> the means to preserve those weaker, less robust individuals who, by nature, are denied horns or the sharp fangs of the beasts of prey with which to wage the struggle for existence. This art of dissimulation reaches its peak in humankind, where deception [...] is so much the rule and the law.[1]

Humanity has occluded its own animality; humans hide from the truth of their bodies and their animal life, keeping "far away from the twists and turns of the bowels, the rapid flow of the blood stream, and the complicated tremblings of the nerve-fibres".[2] A similar motif is at work in the

G. Whiteley (✉)
Stockholm University, Stockholm, Sweden
e-mail: giles.whiteley@english.su.se

R. Hawthorn, J. Miller (eds.), *Animals in Detective Fiction*, Palgrave Studies in Animals and Literature,
https://doi.org/10.1007/978-3-031-09241-1_2

detective fiction of Charles Dickens, for whom the power of dissimulation, identified by Nietzsche as proper to humanity, both explains the criminal and their potential criminal successes: "wearing masks" allows the criminal to potentially escape detection, to pass themselves off as simply another member of the pack. It is such dissimulation that Dickens' Inspector Bucket of *Bleak House* (1852–53) discerns; he "possess[es] an unlimited number of eyes",[3] his gaze permitting him to see what remains hidden to the other characters in the novel. For Nietzsche, this hidden thing is precisely the animal within man, which is "pitiless, greedy, insatiable, and murderous", as if the human being were "hanging in dreams [...] on the back of a tiger" (143). Likewise, Dickens links criminality in both *Bleak House* and his other detective novel, the unfinished *The Mystery of Edwin Drood* (1870), to the signs of the tiger. As such, Dickens' detective fiction operates in relation to discourses of predatory animality.

While research on animals in Dickens has proliferated in recent years,[4] *Edwin Drood* has been virtually ignored by such criticism, as have the ways in which Dickens' contributions to the genre of detective fiction rest upon a series of important allusions to and representations of animals. This chapter follows Dickens' use of animal motifs in his detective fiction, focusing on *Bleak House* and *Edwin Drood*. I zoom in on these two novels in particular because, while neither can be wholly reduced to the genre of detective fiction—both marked by a generic capaciousness and incorporating diverse modes of writing—and while Dickens also wrote elsewhere of the figure of the detective—most notably in his articles for *Household Words* on the real-life figure of Inspector Charles Frederick Field (1805–1874)—it is in *Bleak House* and *Edwin Drood* that he investigates the relationship between animality and detection in most interesting ways. The essay begins by considering how Dickens' treatment of "tigerish" criminals seemingly repeats uncritically an Orientalist discourse, but also subtly deconstructs these tropes. In an analysis informed by Jacques Derrida's linking of the animal, the criminal and the "rogue", I examine the ways in which Dickens' use of animals troubles simple dichotomies which too uncritically pit human beings as figures of law and culture against animals as cyphers of criminality. Far from keeping the figures of the detective and the criminal separate, Dickens' detective fiction instead complicates them. For Dickens, the detective and the criminal are caught in a chiastic structure, the tiger-hunter identifying and identified with their prey. This kind of uncanny doubling of the figures of detective with the criminal and vice versa constitutes an insight that was a formative contribution to the detective genre as a whole and a facet that develops through the dialogue of Dickens' work with evolutionary theory. In this sense, while resting on racist imagery, Dickens' treatment is complex, with animals deployed in order to tease out some of the boundaries separating man from criminal, but also showing that such divisions are untenable. The essay concludes by moving from Dickens' use of felines to considering his treatment of birds. If Dickens' various tigers link the detective and the criminal, then his birds, and rooks in particular, make a broader case, deconstructing the supposed limits that divide the law and those who represent it from those who transgress its boundaries.

Tigers and Rogues

In *The Beast and the Sovereign*, Derrida deconstructs the "accredited oppositional limits" between nature/culture and animal/man.[5] Derrida notes that, alongside beast and sovereign, another figure who finds themselves outside the law is the criminal, and "between sovereign, criminal, and beast" there is "a sort of obscure and fascinating complicity" (17). As such, the criminal is a kind of "rogue". Derrida notes that the meaning of this word gradually extended from its original Elizabethan legal designation for an idle vagrant or vagabond, coming to apply to

> all nonhuman living beings, that is, [...] plants and animals whose behavior appears deviant or perverse. Any wild animal can be called rogue but especially those [...] that behave like ravaging outlaws, violating the customs and conventions, the customary practices, of *their own* community (93).

The "rogue" is the "outlaw", one who stands outside the law, who refuses to abide by what is supposedly "proper". As a typical example of this kind of "rogue" animal, Derrida gives the figure of the tiger. There is of course a kind of "asininity", something itself eminently deconstructible, in this reductive rhetorical gesture of taking one animal as signifier of the "rogue" in general. But Derrida considers the tiger exemplary insofar as it is a lone predator which attacks civilised culture from the margins, picking off the herd. The tiger is an "individual who does not even respect the law of the animal community, of the pack, the horde, of its kind. By its savage or indocile behaviour, it stays or goes away from the society to which it belongs" (19). As rogue, the tiger is both beast and sovereign, sitting at the limits of society, like the criminal.[6] It is significant in this context to note that in both his detective novels, Dickens describes the criminal and their violent behaviour as being "tigerish".

Bleak House follows the legal case of Jarndyce and Jarndyce, a "scarecrow of a suit [that] has, over the course of time, become so complicated, that no man alive knows what it means" (15). This case stands for the Law, which figures in turn for the mid-Victorian British legal system and the politics which underwrite it. Mr. Tulkinghorn, a lawyer whose "calling is the acquisition of secrets, and the holding possession of such power as they give him", is intricately involved with both the case and the Dedlocks, the aristocratic house that he protects (537); he is murdered in his chambers in Lincoln's Inn Fields under a painting of Allegory (692–93). Dickens invites his readers to play detective, not only in trying to divine the identity of Tulkinghorn's murderer, but in seeking to solve the allegorical significance of his death.[7] The event signifies the coming of a new age in which the old British class system gives way to a more democratic order, emblematised by the detective Bucket himself. This character was based on the same Inspector Field of whom Dickens would write in *Household Words*, one of the detectives of the Metropolitan Police, recently established in 1842, marking a new and explicitly "modern" form of criminal investigation.[8] Tulkinghorn's murderer is eventually revealed to have been Lady Dedlock's French maid, Madame Hortense. When finding herself under Tulkinghorn's gaze, Hortense is described as having a "feline personage" which is suspicious, "her eyes looking out at him sideways" (614).[9] From the moment of her first introduction, Hortense is a woman "who would be handsome, but for a certain feline mouth, and general uncomfortable tightness of face, rendering the jaws too eager, and the skull too prominent" (171). The feline is clearly atavistic, with Dickens' reference to the shape of her skull implicated within the discourse of phrenology, the adverbial "too" speaking to an excessiveness, suggesting a predatory bent, one which cannot be domesticated. Similar associations attend the other animals with which she is associated: the vixen (615) and "a very neat She-Wolf imperfectly tamed" (171). Dickens' allusions are obviously misogynistic,[10] so that Hortense is figured as a woman who does not know her place, who is insufficiently submissive—women hide "their tiger's claws inside their glove", as Nietzsche puts it.[11] More broadly, both the wolf and tiger are numbered by Derrida as "rogue" animals.[12] Hortense is such a "rogue", a "domestic" who has turned on her mistress. Unsurprisingly, then, we find that when finally apprehended by Bucket and faced with her crimes, Hortense smiles "with that tigerish expansion of the mouth" that speaks to her guilt (768).[13]

These associations are ones that Dickens would also deploy nearly two decades later in *The Mystery of Edwin Drood*. Written after the Indian Rebellion of 1857, in this novel Dickens mobilises the tiger as a specific signifier of the Orient, playing upon stereotypical connotations for mid-Victorian readers. Interpreting this novel is a notorious exercise in speculation given that Dickens died before its completion, but it turns on the eponymous Edwin's disappearance and presumed death. The chief suspect is John Jasper, his uncle, a respectable choirmaster by profession, but addicted to opium. Beyond Jasper, another suspect is Neville Landless who has recently arrived in Cloisterham from Ceylon (today, Sri Lanka).[14] When Mr. Crisparkle first meets Neville and his sister, the narrator notes they were

> both very dark [...] in colour; [...] something untamed about them both; a certain air upon them of hunter and huntress; yet withal a certain air of being the objects of the chase, rather than the followers [...]; fierce of look; an indefinable kind of pause [...] on their whole expression [...] which might be equally likened to the pause before a crouch or a bound (42).

Neville is quick to anger, and he associates this with his upbringing: "I have been brought up among [...] an inferior race, and I may easily have contracted some affinity with them. Sometimes, I don't know but that it may be a drop of what is tigerish in their blood" (47). He is sensitive to racial allusions, and when Edwin calls him "a common black fellow" (61), Neville makes to attack him, before Jasper intervenes. Moments later, Neville meets Mr. Crisparkle, telling him that Edwin had "heated that tigerish blood I told you of" (62). Jasper, although not privy to these discussions, concurs with the allusion, noting himself separately that "there is something of the tiger in his dark blood" (63).

Jasper uses this image of Neville's violence as indissolubly linked to the tigerishness of his blood to lay the blame for Edwin's disappearance on him, and the fact that Jasper makes this same allusion to "tigerish blood" independently of Neville is unsurprising given the context of his upbringing in Ceylon. During the Victorian period, the tiger was inextricably bound up with the narratives of Empire. As Elsie Cloete argues, across the period, "the Mughal symbol of the tiger as an animal of power and potency" gradually "morphed [...] into one of 'Oriental ferocity and motiveless violence'".[15] This was particularly true in the wake of the Indian Rebellion, in which "the tiger became a defining image of the conflict, giving rise to a symbolism that operated through a tense juxtaposition of simile and ethnography".[16] In speaking of Neville's "tigerish blood", Jasper seeks to capitalise on "a provincialism that reveals itself as pure racism".[17] Of course, such a provincialism is also at work in Dickens' own use of the tiger as a metonym for the East, as much as his characters'. As such, Dickens' writing ends up necessarily somewhat complicit in the very structures he is working to criticise.

Animal Expressions

Dickens suggests that the signs of the tiger betray a savage violence that may erupt at any time. Tigers are outsiders whose "passion" is opposed to domestic, British values, law and social order. The famous "Tipu's Tiger" (Fig. 1) (c. 1793), a musical organ formed to depict a tiger mauling a European figure, images the Indian tiger as a return of the colonial repressed. As the nineteenth century progressed, the tiger came to signify "the enduring spirit of India that the British felt they had failed to subjugate. No matter how many successful campaigns the British had waged [...], some basic fear of India continued to haunt" their imagination.[18]

Fig. 1 Tipu's Tiger

The parochial panic that "tigerish blood" precipitates in Cloisterham's Home Counties is hardly surprising in this context, and a similar panic is also present in Wilkie Collins' contemporaneous detective novel, *The Moonstone* (1868), where the Indians are described by Mr. Murthwaite as "men who will wait their opportunity with the patience of cats, and will use it with the ferocity of tigers".[19] But if Dickens' novel repeats a well-established pattern of Orientalist discourse somewhat uncritically, he also begins to challenge these ideas by linking the detective and the criminal together. Placing the novel into dialogue with Dickens' interest in evolutionary biology, we begin to see the ways in which his detective fiction plays out some of the themes of that essential nineteenth-century colonial rite of passage: the tiger hunt.

The tiger hunt was "one of the most violent and visible spectacles of political authority" exercised by the British in colonial India, a complicated ceremonial event in which "predatory metaphors of rule" became concretised in the form of "hunting practices".[20] As the century wore on, and particularly in the wake of the 1857 "Mutiny", "the tiger-hunter himself became somewhat of a stock type" in the fiction of the period.[21] But it was also the case that these later nineteenth-century tiger-hunters were not wholly divorced from their quarry, and began to identify with their prey in a chiastic structure that is also present in Victorian detective fiction. In *Bleak House*, Inspector Bucket is a kind of domesticated doppelgänger of the criminal Hortense: a "Tom Cat" rather than a tiger (373). And as the genre of detective fiction developed in the wake of Dickens, such uncanny identifications became more common. Take, for instance, Arthur Conan Doyle's "The Empty House" (1904), in which the criminal, Colonel Sebastian Moran, once of Her Majesty's Indian Army, is said to look "like a tiger himself", but where Holmes too is likened to a tiger.[22] This chiastic structure is also present in the figures of Neville and Helena in *Edwin Drood*, who have the air of both "hunter" and "of being the objects of the chase" (42). A predator in one context may be prey in another.

The identificatory structure linking tiger and hunter was accentuated by evolutionary theory. As Heather Schell puts it, "the hunters' sense of kinship with tigers was infused with a nascent conviction that masculinity was itself predatory, and that predators made an important contribution to natural selection" (230).[23] In *The Descent of Man* (1871), Darwin made the link between the predator and the prey, with whom they began to identify. Speaking of tigers, Darwin notes that both fed on the native population and were independent or self-sufficient.[24] "The tiger was a solitary hunter", Schell comments, and "British Indian Shikari [Urdu: hunter] also perceived themselves as solitary hunters" (239–40). For Darwin, the tiger stood outside the pack as a lone figure, for while "no doubt a tiger [...] feels sympathy for the sufferings of its own young", it did not demonstrate any discernible sympathy "for any other animal" (130). This kind of sympathy, "the mutual love of the members of the same community", has its evolutionary benefits, Darwin argues, helping the community to grow (130), so a lack of sympathy speaks to the tiger's status as "rogue". Developing the point, Derrida argues that natural selection has traditionally needed a helping hand from culture in "roguing out" the rogue. For his definition of the "rogue",[25] Derrida cites Darwin's discussion of individuals "that deviate from the proper standard" in *The Origin of Species* (1859)[26]: the community, founded on an idea of sympathy that "rogues" such as the tiger do not feel, requires that such figures be ostracised from the pack.

While *Bleak House* was published before *The Origin of Species*, Dickens became keenly interested in evolutionary theory in the decade before his death, as Gillian Beer and George Levine have shown.[27] *Edwin Drood*, for its part, was written and conceived explicitly in an evolutionary context.[28] Its plot rests upon seeing and interpreting the traces of the past which have conspired to create the present events. Neville is not simply "tigerish"; rather, the tiger is also known through the expression of his emotions. When he comes across Mr. Crisparkle after his fight with Edwin, Neville's body tells its tale unconsciously:

> "Mr. Neville", rejoins the Minor Canon, mildly, but firmly: "I request you not to speak to me with that clenched right hand. Unclench it, if you please."
>
> "He goaded me, sir", pursues the young man, instantly obeying, "beyond my power of endurance. [...] In short, sir", with an irrepressible outburst, "in the passion into which he lashed me, I would have cut him down if I could, and I tried to do it".
>
> "You have clenched that hand again", is Mr. Crisparkle's quiet commentary (62).

Mr. Crisparkle is a figure associated with the Church and thereby the law, acting as guardian to Neville while in Cloisterham, but even he is unable to tame him: his initial correction lasts less than the time taken for Neville to reply. The tiger in him is "irrepressible".

The animal within man is seen in Neville's clenched fist, that "most incorporate tool", as Samuel Butler put it, one that retains "something of the non-ego about it".[29] It allows the reader to see through the eyes of Mr. Crisparkle, to act as detectives in glimpsing the signs of the tiger. The passage anticipates Darwin in *The Expression of Emotions on Man and Animal* (1872). In moments of anger,

> the body is commonly held erect ready for instant action [...]. Such gestures as the raising of the arms, with the fists clenched, as if to strike the offender, are common. Few men in a great passion [...] can resist acting as if they intended to strike or push the man violently away.[30]

And while *Edwin Drood* was published in 1870, predating the publication of *The Expression of Emotions*, Dickens and Darwin had a common source for their respective discussions of the language of "passionate" fists in Sir Charles Bell's *The Anatomy and Philosophy of Expression* (1806). As Jonathan Smith has shown, Darwin's text was conceived in part as an attempt to "answer" Bell by showing that expression is not divinely given but the product of evolutionary development.[31] In this sense, reading the expressions of man becomes a kind of detective work, the naturalist-cum-inspector discerning the signs of our animal ancestors. Indeed, the similarities between these two new kinds of mid-Victorian specialists were not lost on proponents of evolutionary theory. The zoologist Andrew Wilson pointed out in 1879 that "the naturalist of to-day bears a certain likeness to the detective officer".[32] Darwin even portrays himself as a detective in *The Expression of Emotions*, discussing the "guilt" of his two-year old son, revealed by his blushing which "led to the detection of his little crime" (261). Bell, for his part, had discussed clenched fists, and Dickens owned the third edition of his book.[33] And that Darwin himself thought Dickens' texts anticipated his own evolutionary theory is shown by his approving quotation of *Oliver Twist* (1837–39), where people are seen "snarling with their teeth [...] like wild beasts".[34] It seems important in this context that Darwin's allusion to Dickens in *The Expression of Emotions* immediately followed a citation of Bell. Like Dickens, Darwin recognises that such "expressions" can be read, revealing the animal signs that the human being seeks to keep hidden behind their masks.

Animals before the Law

Taken in context, the image of the tiger diagnoses Dickens' criminals qua criminals. They are foreign and lack that "sympathy" which makes them trustworthy, and which would make them sympathetic themselves. Dickens' use of the image of the tiger risks a kind of schematism symptomatic of the racist connotations he plays upon: Criminals are criminals because they are tigers and their tigerishness speaks to the animal within man. Dickens appears to be saying that we can detect criminals by paying attention to the "tells" they give, the ways in which they act like, or resemble in some manner, certain animals, and particularly "rogue" animals. In *Bleak House*, Esther describes Mr. Vholes, Richard's lawyer, as a parasite feeding off the body of his client: "So slow, so eager, so bloodless and gaunt, I felt as if Richard were wasting away beneath the eyes of this adviser and there were something of the vampire in him" (854). The same analogy attends Mr. Krook, the rag merchant who is landlord to Nemo and Miss Flite, and who later dies of spontaneous combustion. When he and Tulkinghorn find Nemo dead in his room, Krook's "lean hands spread out above the body like a vampire's wings" (152). Likewise, both Vholes and Krook are associated with the fox, noted for its "slyness" (218, 582), and thereby Hortense. The moral lesson of Dickens' detective fiction seems relatively clear-cut in this context: if we all have an element of the tiger in us, better to be the Tom Cat Bucket, employing those skills in the name of social good. In this sense, it is only the law (of the pack, community), and Bucket as an instrument of its justice, which prevents us all from regressing to the same "tigerish" state as a Hortense, from becoming one of Nietzsche's "beasts of prey".

But attractive as it may be, Dickens does not make it quite so easy for the reader or his detectives. Here it is instructive to compare the cats and the birds of his detective fiction. Take another of *Bleak House*'s tigers, the "large grey cat", Lady Jane, who belongs to Krook:

> The cat leaped down and ripped at a bundle of rags with her tigerish claws, with a sound that it set my teeth on edge to hear.
>
> "She'd do as much for anyone I was to set her on", said the old man (64).

Sadistic and predatory, Lady Jane is an ill-fitting pet, just as Hortense is an ill-fitting domestic. She is fixated upon Miss Flite's birds, so "sly, and full of malice" that she "half believe[s], sometimes, that she is no cat, but the wolf of the old saying" (67). The allusion to the idiom of the wolf at the door makes Lady Jane an image of hunger: an allegory of the insatiability of animal desire, where the door represents the safety of the domestic interior. Miss Flite's birds are also named after allegorical figures (217, 853), kept caged until the Judge determines the suit of Jarndyce and Jarndyce (524, 904). These birds in their turn allegorise the case, so that their incarceration stands for the bureaucratic labyrinths of the mid-Victorian legal system, and the predation of Lady Jane stands for the Law itself. Derrida reminds us of the complex relationship that links the rogue, the criminal, the beast and the sovereign, so that "being-outside-the-law" can "take the form of the Law".[35] As a "rogue", Lady Jane cannot be trusted; like her owner, the crooked Krook, and like the Law, she feeds off the society she is supposed to protect. Vholes too has an "official cat who is patiently watching a mouse's hole", simultaneously both a doppelgänger of Lady Jane and a simulacrum of its owner (576). More significantly, perhaps, Lady Jane's "natural cruelty is sharpened by her jealous fear of their [the birds'] regaining their liberty" (67), so that her inner tiger will be released when the birds are. This is exactly what Derrida classifies as the "freedom" of sovereignty, that state of "being-outside-the-law" that is also characteristic of the rogue. When Nemo is found dead, the surgeon tells Krook that Lady Jane cannot be allowed to remain with the cadaver, "and she goes furtively down stairs, winding her lithe tail and licking her lips", before Tulkinghorn "goes home to Allegory" (157). The immediate allusion to the painting invites us to read Lady Jane's "tigerishness" as a kind of cannibalism and the passage allegorically. It recalls the character of the Marquis St. Evrémonde in *A Tale of Two Cities* (1859), who stands for the aristocratic *ancien régime*, a predatory figure feeding on the lower classes, and who is spoken of as a tiger.[36] Lady Jane is hoping to feed off the body of her (master's) lodger, but the allusion to "Allegory" in the passage also invites us to read the Law as an institution itself "naturally cruel". There is a kind of sadism in the Law, as Derrida has argued with reference to another Marquis of the French Revolution: de Sade.[37] Such "cruelty" is allied with the spectacular and with the "spectacle" as a simultaneous expression of disciplinarity and *jouissance*. It is partly in this sense that Miss Flite speaks of the "cruel attraction" of the courtroom (523). The Law captures and captivates so many of the characters in the novel, who become ensnared in its "charmed gaze".

In *Edwin Drood*, too, it is the birds as much as the tigers that serve to remind us of the precariousness of any attempts at delineating the boundaries between those who stand outside the Law and those who represent it. Dickens himself was the owner of a raven named Grip, who would be immortalised in the form of the protagonist's talking companion in *Barnaby Rudge* (1841), and who would, in turn, inspire Edgar Allen Poe's "The Raven" (1845). But if Dickens was partial to his raven, other types of corvidae came off less favourably in his work. Unlike Miss Flite's unclassified birds, those in *Edwin Drood* are clearly named as rooks, and as Jane Vogel notes, "rooks congregate where Law and 'letter' rule" in Dickens (324). In *Bleak House*, the rook is associated with the Law through the figure of Mr. Tulkinghorn, whose "black figure" appears "like a larger species of rook" (175). In *Edwin Drood*, by contrast, corvidae are explicitly associated with the Church, with "hoarse rooks hovering about the Cathedral tower, its hoarser and less distinct rooks in the stalls far beneath" (13), and the use of "rook" as slang for a member of the clergy is dated by the *OED* to 1859 ("rook" n.[1], 2d). Dickens had previously drawn the association between the rook and "the stately, grey Cathedral" of *David Copperfield* (1849–50), where the birds are associated with Dr. Strong, David's teacher at school.[38] The school teaches discipline, linking education to the Law and the Church. Unsurprisingly, then, we find that the "clerical bird" (3) also has a panoptic function in *Edwin Drood*. The "cawing of the rooks" accompanies the "shadow of the Cathedral" (38) where Jasper works, and from this vantage point, they oversee most of the events of the narrative. For instance, when Edwin and Rosa decide to separate, "the rooks hovered above them with hoarse cries, darker splashes in the darkening air" (120).

As panoptic animals, the rooks of Cloisterham ally with the figure of the detective in Dickens' fiction: their unrelenting gaze recalls Inspector Bucket's eyes in *Bleak House* (335). But in *Edwin Drood*, the rooks also miss the moment of the murder or the disappearance of Edwin himself, registered only metonymically through their empty nests: "The darkness is augmented and confused, by flying dust from the earth, dry twigs from the trees, and great ragged fragments from the rooks' nests up in the tower" (131). The absence of the rooks suggests a limit to the power of the Law and its panoptic dreams. Dickens means the allusion to recall *Macbeth* (the chapter is titled "When shall these three meet again?", taking in the play's opening line), in which "rooks" are said to have "brought forth / The secret'st man of blood" and in which "the crow / Makes wing to the rooky wood", a phrase alluded to earlier in *Edwin Drood*.[39] The allusions associate the rook with death and uncertain augury, and Dickens suggests that neither the Law nor any detective is able to see through such a darkness. Nor is the rook a paragon of godliness or virtue, as the association between the crow and Tulkinghorn's house in Lincoln's Inn Fields in *Bleak House* makes clear (145, 147). Dickens no doubt meant to play upon the fact that, as a verb, "to rook" means to steal or swindle (*OED*: "rook", v.[2], 1a); as such, the rook implies the crook. The Law, which is meant to protect the weak, ends up preying upon them, like the tiger or vampire. The Law, for all its pretensions, becomes cannibalistic, devouring those it was meant to serve.

What Dickens' detective fiction drives towards, then, is the deconstruction of the boundaries between those who stand outside the Law and those who represent it. The detective becomes allied with the criminal in a chiastic structure, just as the hunter came to identify with the tiger. Even those who stand for the Law, such as Tulkinghorn or the rooks, are revealed to be another kind of "rogue". We may make the point one last time by reference to another pair of cats from *Edwin Drood*. Edwin is betrothed to Rosa Bud, who he nicknames "Pussy", the "pet pupil of the Nuns' House" (14), her school, and who Edwin himself has become used to "petting" (64). She is shy and associated with the domestic, the quintessential English Rose, as her name implies. But at the same time, Pussy's "innocence" (116) does not preclude a different image of petting, as Edwin's "pet name" takes in the coarse slang, a point which Dickens plays upon in an awkward exchange with Mr. Grewgious, where Edwin blushes in a moment where his (animal) expression gives him away (90). Later in the novel, Rosa deliberately casts off the name "Pussy", claiming thereby a measure of independence, but earlier in the text, she appears satisfied to play the role of coquette. This is seen in an oblique allusion where Pussy exclaims how "it *is* so absurd to have girls and servants scuttling about after one, like mice in the wainscot" (15–16), somewhat disingenuously given that she is pleased by the attention she gets at the Nun's House. In Anna Laetitia Barbauld's *Lessons for Children* (1778–79), the narrator tells a story about "Puss" who "smells the mice. They are making a noise behind the wainscot", and it seems likely that Dickens was alluding to this passage.[40] Perhaps most importantly for our purposes, while Barbauld's "Puss" is a pet, domesticated and made familiar by her shortened name, in practice, she is a cold predator, catching the mice, toying with them before killing them, so that the narrator admonishes Puss "not [to] be so cruel" (27).

In this context, Dickens' allusion to Barbauld invites comparison between Pussy as an image of the perfect English Rose and the Princess Puffer, Jasper's opium dealer in Blue Gates Fields. She is another "rogue", linked to criminal elements, as well as to a kind of decadence which is "foreign", with opium associated with the Chinese immigrants of the area, and simultaneously implying the East, those "routes of financial exploitation and moral contamination" that link London to the world beyond.[41] When Jasper visits the Princess Puffer in the den in the final extant chapter of the novel, "she lays her hand upon his chest, and moves him slightly to and fro, as a cat might stimulate a half-slain mouse" (209). Significantly, it is a moment when Princess Puffer is herself playing detective, for she is seeking to elicit from Jasper a confession for the murder she suspects him of committing, and a moment later, she repeats this "catlike action" (210). These feline allusions associate Princess Puffer with the "rogue" Neville and his "tigerish blood", as well as associating the "rogue" once more not only with the figure of the criminal but also with that of the detective. But perhaps most interestingly, the feline here also necessarily implicates Pussy by the association, so that the image of the English Rose becomes a kind of doppelgänger of the Princess Puffer, and vice versa, the one ghosting the other according to a kind of "an *unheimlich*, uncanny, reciprocal haunting".[42] That which was most domestic, pure and cloistered becomes tied to the foreign, the "rogue" and the animal within man.

Ultimately, it is this kind of haunting which underwrites Dickens' detective stories. In both *Bleak House* and *Edwin Drood*, the criminal and the detective, the rogue and the representatives of the law, become caught in a chiastic structure of mutual co-implication. If Dickens initially seems to be associating the animality of his criminals with their "foreignness", then he is also aware that it is precisely these kinds of divisions which detective fiction reveals to be untenable.

Notes

1. Nietzsche, *Birth of Tragedy*, 142.
2. Nietzsche, *Birth of Tragedy*, 142.
3. Dickens, *Bleak House*, 335.
4. For Dickens's dogs, for instance, see Kreilkamp, "Dying Like a Dog"; Moore, "Beastly Criminals"; Gray, *The Dog in the Dickensian Imagination*;

McDonell, "Bull's-eye". More broadly, for a sociological perspective on Dickens and animals, see Morrison, "Dickens, *Household Words*, and the Smithfield Controversy". Olson, *Criminals as Animals*, 251–74 is interested in similar issues as I am in this essay, but differs in focus, not considering the question of detective fiction as a genre, dealing only peripherally with *Bleak House* and not considering *Edwin Drood.*

5. Derrida, *Beast and the Sovereign*, 15.
6. See Krell, *Derrida and Our Animal Others*, 10.
7. Allegory, in this sense, always being catachresis, in a motif that links to Nietzsche on the human power of willed dissimulation. On Dickens and allegory, in an analysis informed by Walter Benjamin, see Tambling, *Dickens' Novels as Poetry*, and more broadly, see Vogel, *Allegory in Dickens.*
8. Dickens' writing on Field dates from the period of *Bleak House.* He published four pieces on Field, "A Detective Policy Party" in two parts, "Three 'Detective' Anecdotes" and "On Duty with Inspector Field" in *Household Words* on 27 July, 18 August, 14 September 1850 and 14 June 1851, respectively. With William Henry Wills, he also co-wrote an essay on "The Metropolitan Protectives", published in *Household Words* on 26 April 1851. On Field, Bucket and these articles, see Collins, *Dickens and Crime*, 204–209.
9. Of particular significance given that Dickens was a great dog lover, his texts replete with positive canine imagery: see Gray, *The Dog in the Dickensian Imagination.*
10. Recall Derrida: "the she-wolf [is] often a symbol of sexuality or even of sexual debauchery" (*Beast and the Sovereign*, 9).
11. Nietzsche, *Beyond Good and Evil*, 129.
12. Derrida, *Beast and the Sovereign*, 64.
13. This fits into a wider pattern in Dickens' work. As a child, Dickens read in the *Portfolio* of tigers which, even when fed, "still continue[d] the carnage, [...] destroying one [animal] after another, [...] as if it increased, rather than satiated his voracious appetite" (Stone, *Night Side*, 132). Later, in *Great Expectations* (1860–61), we find that just before Dolge Orlick tries to murder Pip, "his mouth snarl[ed] like a tiger's" (388).
14. Dickens, *Edwin Drood*, 46.
15. Cloete, "Tigers, Humans and *Animots*", 315.
16. Miller, "Rebellious Tigers", 480.
17. Frank, *Victorian Detective Fiction,* 109.
18. Mukerjee in Cloete, "Tigers, Humans and *Animots*", 315–16.
19. Collins, *Moonstone*, 72. *The Moonstone* was a key inter-text for the writing of *Edwin Drood.* Collins and Dickens were firm friends, and they had recently collaborated on *No Thoroughfare* (1867); like *The Moonstone* and

Edwin Drood, this play turns on a character's use of drugs and on their subsequent inability to recollect events.

20. Pandian, "Predatory Care", 80.
21. Schell ,"Tiger Tales", 229. For more on the ways in which the tiger was coded following the 1857 Rebellion, see Miller, "Rebellious Tigers".
22. Conan Doyle, *Return of Sherlock Holmes*, 22, 20. On this, see Schell, "Tiger Tales", 240.
23. Although interestingly, "the tiger, one of the most beautiful animals in the world", is one in which "the sexes of which cannot be distinguished by colour, even by the dealers in wild beasts", Darwin remarks (2004: 609). Note that Neville, the "tiger" of *Edwin Drood*, has a twin sister in Helena, his doppelgänger for whom he may be mistaken.
24. Darwin, *Descent of Man*, 65.
25. Derrida, *Beast and the Sovereign*, 93.
26. Darwin, *Origin of Species*, 28.
27. See Beer, *Darwin's Plots*, 40–43, and Levine, *Darwin and the Novelists*, 119–76.
28. See Frank, *Victorian Detective Fiction*, 99–130.
29. Butler, *Notebooks*, 18.
30. Darwin, *Expression of the Emotions*, 236.
31. Smith, *Charles Darwin*, 186–98.
32. Wilson, "Clews", 14.
33. Bell, *Anatomy and Philosophy of Expression*, 220, and see Stonehouse, *Catalogue*, 11. Dickens met Bell in June 1841 (*Letters*, 309). For some of the ways in which Bell might have influenced Dickens in his writing of "expression and gesture", see McMaster, *Dickens the Designer*, 51–57.
34. Darwin, *Expression of the Emotions*, 238–39; quoting Dickens, *Oliver Twist*, 40.
35. Derrida, *Beast and the Sovereign*, 17.
36. Dickens, *Tale of Two Cities*, 120. There is another important connection between the tigers of *A Tale of Two Cities* and *Bleak House*. Dickens' readers would also have associated Hortense's tigerishness with her French nationality. Such connotations are not lost on Esther, who sees Hortense as "some woman from the streets of Paris in the reign of terror" (339–40). Associating Hortense with the Terror is significant: a time when humans turned on one another, and where the law itself became a form of terrorism.
37. Derrida, *Death Penalty*, 159–70.
38. Dickens, *David Copperfield*, 183, 219.
39. Shakespeare, *Macbeth*, 3.4.123–24, 3.2.51–52. The allusion occurs in the opening paragraph of chapter two of *Edwin Drood* where the bird "wings his way homeward" (3). Dickens replaced "crow" with "rook" when he revised the manuscript for publication.

40. Barbauld, *Lessons for Children*, 26. I have discussed in more detail the evidence for, and significance of, Dickens' allusion to Barbauld elsewhere: see Whiteley, "Oblique Allusion to Barbauld".
41. Agathocleous, *Urban Realism*, 118.
42. Derrida, *Beast and the Sovereign*, 17. I discuss the "haunting" quality of the novel in more detail in Whiteley, *Aesthetics of Space*, 112–14.

Works Cited

Agathocleous, Tanya. *Urban Realism and the Cosmopolitan Imagination in the Nineteenth Century*. Cambridge: Cambridge University Press, 2011.

Barbauld, Anna Laetitia. *Lessons for Children*. New edn. London: Baldwin and Cradock, 1830.

Beer, Gillian. *Darwin's Plots: Evolutionary Narrative in Darwin, George Eliot and Nineteenth Century Fiction*. 2nd edn. Cambridge: Cambridge University Press, 2000.

Bell, Charles. *The Anatomy and Philosophy of Expression*. 3rd edn. London: John Murray, 1844.

Butler, Samuel. *The Notebooks of Samuel Butler*. Edited by Henry Festing Jones. New York: Dutton, 1917.

Cloete, Elsie. "Tigers, Humans and *Animots*", *Journal of Literary Studies / Tydskrif vir Literatuurwetenksap* 23, no. 3 (2007): 314–333.

Collins, Philip. *Dickens and Crime*. 3rd edn. London: Macmillan, 1994.

Collins, Wilkie. *The Moonstone*. Edited by John Sutherland. Oxford: Oxford University Press, 2008.

Conan Doyle, Arthur. *The Return of Sherlock Holmes*. Harmondsworth: Penguin, 2008.

Darwin, Charles. *On the Origin of Species*. Edited by Gillian Beer. Oxford: Oxford University Press, 2008.

Darwin, Charles. *The Descent of Man: Selection in Relation to Sex*. Edited by James Moore and Adrian Desmond. Harmondsworth: Penguin, 2004.

Darwin, Charles. *The Expression of the Emotions in Man and Animals*. Edited by Paul Erkman. Oxford: Oxford University Press, 1998.

Derrida, Jacques. *Rogues: Two Essays on Reason*. Translated by Pascale-Anne Brault and Michael Naas. Stanford: Stanford University Press, 2004.

Derrida, Jacques. *The Beast and the Sovereign, Volume 1*. Translated by Geoffrey Bennington. Chicago: University of Chicago Press, 2009.

Derrida, Jacques. *The Death Penalty, Volume 1*. Translated by Peggy Kamuf. Chicago: University of Chicago Press, 2014.

Dickens, Charles. *A Tale of Two Cities*. Edited by Andrew Sanders. Oxford: Oxford University Press, 2008a.

Dickens, Charles. *Bleak House*. Edited by Stephen Gill. Oxford: Oxford University Press, 2008b.

Dickens, Charles. *David Copperfield*. Edited by Andrew Sanders. Oxford: Oxford University Press, 2008c.

Dickens, Charles. *Great Expectations*. Edited by Robert Douglas-Fairhurst. Oxford: Oxford University Press, 2008d.

Dickens, Charles. *Oliver Twist*. Edited by Kathleen Tillotson. Oxford: Oxford University Press, 2008e.

Dickens, Charles. *The Letters of Charles Dickens: Volume Two, 1840–1841*. Edited by Madeline House & Graham Storey. Oxford: Clarendon, 1969.

Dickens, Charles. *The Mystery of Edwin Drood*. Edited by Margaret Cardwell. Oxford: Oxford University Press, 2009.

Frank, Lawrence. *Victorian Detective Fiction and the Nature of Evidence*. London: Palgrave Macmillan, 2003.

Gray, Beryl. *The Dog in the Dickensian Imagination*. London: Routledge, 2014.

Kreilkamp, Ivan. "Dying Like a Dog in Great Expectations". In *Victorian Animal Dreams*, edited by Deborah Denenholz Morse and Martin A. Danahay, 81–94. London: Ashgate, 2007.

Krell, David Farrell. *Derrida and Our Animal Others*. Bloomington: Indiana University Press, 2013.

Levine, George. *Darwin and the Novelists: Patterns of Science in Victorian Fiction*. Cambridge, Mass.: Harvard University Press, 1988.

McDonell, Jennifer. "Bull's-eye, Agency, and the Species Divide in *Oliver Twist*: a Cur's-Eye View." In *Animals in Victorian Literature and Culture*, edited by Laurence W. Mazzeno and Ronald D. Morrison. London: Palgrave Macmillan, 2017.

McMaster, Juliet. *Dickens the Designer*. London: Macmillan, 1987.

Miller, John. "Rebellious Tigers, a Patriotic Elephant and an Urdu-Speaking Cockatoo: Animals in 'Mutiny' Fiction", *Journal of Victorian Culture*, 17, no. 4 (2012): 480–91.

Moore, Grace. "Beastly Criminals and Criminal Beasts: Stray Women and Stray Dogs in *Oliver Twist*." In *Victorian Animal Dreams*, edited by Deborah Denenholz Morse and Martin A. Danahay, 201–14. London: Asghate. 2007.

Morrison, Ronald D. "Dickens, *Household Words*, and the Smithfield Controversy at the Time of the Great Exhibition." In *Animals in Victorian Literature and Culture*. Ed. Laurence W. Mazzeno and Ronald D. Morrison. London: Palgrave Macmillan, 2017.

Nietzsche, Friedrich. *Beyond Good and Evil*. Translated by Judith Norman. Cambridge: Cambridge University Press, 2002.

Nietzsche, Friedrich. *The Birth of Tragedy and Other Writings*. Translated by Ronald Speirs. Cambridge: Cambridge University Press, 1999.

Olson, Greta. *Criminals as Animals from Shakespeare to Lombroso*. Berlin: De Gruyter, 2013.

Pandian, Anand S. "Predatory Care: the Imperial Hunt in Mughal and British India", *Journal of Historical Sociology* 14, no. 1 (2001): 79–107.

Schell, Heather. "Tiger Tales." In *Victorian Animal Dreams*, edited by Deborah Denenholz Morse and Martin A. Danahay, 229–48. London: Asghate, 2007.

Shakespeare, William. *Macbeth*. Edited by Sandra Clark and Pamela Mason. London: Bloomsbury, 2015.

Smith, Jonathan. *Charles Darwin and Victorian Visual Culture*. Cambridge: Cambridge University Press, 2006.

Stone, Harry. *The Night Side of Dickens: Cannibalism, Passion, Necessity*. Ohio: Ohio State University Press, 1994.

Stonehouse, J.H., ed. *Catalogue of the Library of Charles Dickens at Gad's-hill*. London: Piccadilly Fountain Press, 1935.

Tambling, Jeremy. *Dickens' Novels as Poetry: Allegory and Literature of the City*. London: Routledge, 2015.

Thomas, Ronald R. *Detective Fiction and the Rise of Forensic Science*. Cambridge: Cambridge University Press, 1999.

Vogel, Jane. *Allegory in Dickens*. Alabama: University of Alabama Press, 1977.

Whiteley, Giles. "An Oblique Allusion to Barbauld in *The Mystery Of Edwin Drood*", *Dickens Quarterly* 34, no. 2 (2017): 172–75.

Whiteley, Giles. *The Aesthetics of Space in Nineteenth-Century British Literature, 1843–1907*. Edinburgh: Edinburgh University Press, 2020.

Wilson, Andrew. "Clews in Natural History", *Popular Science Monthly* 15 (May 1879): 14–29.

Quantum Entanglements in Arthur Conan Doyle's *The Hound of the Baskervilles*

Adrian Tait

> A hound it was, an enormous coal-black hound, but not such a hound as mortal eyes have ever seen. Fire burst from its open mouth, its eyes glowed with a smouldering gaze, its muzzle and hackles and dewlap were outlined in flickering flame. Never in the delirious dream of a disordered brain could anything more savage, more appalling, more hellish, be conceived than that dark form and savage face which broke upon us out of the wall of fog.[1]

First published in 1901–2, *The Hound of the Baskervilles* is widely regarded as Holmes' most famous and perhaps even best case. It is also Arthur Conan Doyle's "most dog-centred story".[2] For those who read it with an interest in human–animal relations, however, the novella may in the end feel like something of a disappointment. If it at first seems that Doyle has presented a narrative of resistance to the exploitation of the non-human animal by the human—a story in which, perhaps, the animal possesses independence and agency—its denouement reveals the hound to be another, only still more egregious instance of that exploitation: In spite of

A. Tait (✉)
Independent Scholar, Bath, UK

R. Hawthorn, J. Miller (eds.), *Animals in Detective Fiction*, Palgrave Studies in Animals and Literature,
https://doi.org/10.1007/978-3-031-09241-1_3

its "hellish" appearance (151), the "gigantic hound" (20) is simply a phosphorus-painted cross between bloodhound and mastiff, trained as a "cunning device" (157) in "an all too material plan to acquire Baskerville Hall".[3]

Yet the Baskerville hound remains one of the most memorable inventions in Doyle's oeuvre, at once cultural, experiential, biological, and psychical in its construction. One reason for its enduring fascination lies with the narrative's clever construction, which repeatedly defers the final, fateful encounter between hound and Holmes, even as it builds up the reader's sense of what the hound is or might be: Long before the hound bursts from the fog and onto the page, Doyle's narrative has positioned the hound at the intersection of a number of different but shared societal concerns. For example, Doyle's depiction of a ferocious hound spoke directly to contemporary concerns about the dog's place in society. At a moment when society sought to impose what Philip Howell calls "a kind of *cultural* domestication of the dog", Doyle's hound asserted the contrary power of the animal to upset the balance of power between human and non-human.[4] Doyle's depiction also spoke to anxieties about degenerate "reversions to the primitive", even of "regression into vestigial animal states", then current anxieties that are clearly reflected in the titles of two papers—"Some Freaks of Atavism" and "Do We Progress?"—mentioned early in the novella (6).[5] Furthermore, Doyle's "terrible creature" (152) invoked more ancient, even occult fears. The novella's opening discussions turn on the question of whether this "huge black creature" (161) is supernatural in origin; the possibility that it is in fact a "hell-hound", and one of the "fiends of the pit" (96), surfaces repeatedly. More specifically, Doyle's depiction of the hound taps into an extensive body of folklore relating to spectral hounds: it draws heavily on myths of the "Black Dog", an "ancient reminder of the link between the living and the dead", and, in particular, on the legends of "the Wyst or Wish Hounds [that] are part of the very fabric" of Dartmoor's folklore.[6]

Doyle's "fictional account of the black dog legend" also suggests other reasons why, as Christopher Clausen suggests, the hound may be "the most powerful figure of horror in all the literature of crime".[7] "The early chapters of the book are heavy with references to nightmare [and] madness", Clausen notes, and Doyle's "apparition" appears to threaten "the order of the rational mind itself".[8] Man is the reasoning animal, *animal rationale*, and it is reason which (supposedly) sets him apart; yet Holmes, who is the apparent embodiment of a reasoning machine, is so caught up with the thrill of his own hunt that he imperils the very person it is his job to protect. That he should be so preoccupied with the hunt reflects the doublings of the narrative, in which Holmes is himself a hound after the scent, even as his life is imperilled by a hound. As Holmes and the beast are themselves increasingly aligned—the one likened to a bloodhound, the other itself a bloodhound cross—their intimate entanglement throws "into radical question the schema of the human", reopening the question of what is truly meant by "the animal": "[s]pecies-blurring" itself becomes "a source of anxiety".[9]

Thus, and whilst the character of Holmes appears to anticipate a transhumanist intensification of the sovereign human individual, predicated (Cary Wolfe notes) on "rationality, autonomy, and agency", the novella suggests a contrary, posthumanist "decentring of the human" inextricably linked to what Francis O'Gorman calls the "leakage between the world of man and animals".[10] Yet this "leakage" is itself inseparable from the world in which it occurs. As Neil Pemberton notes, the doubling of Holmes and hound is grounded in their mutual "immersion and physical isolation within the Dartmoor landscape".[11] That landscape is itself much more than background or backdrop. Rather, it is an unquiet presence at work within the novella, "inextricable from the text's inner workings".[12] "Always there was this feeling of an unseen force", notes Watson, "a fine net [...] holding us so lightly that it was only at some supreme moment that one realised that one was indeed entangled in its meshes" (121). As Watson's narrative emphasizes, he and Holmes and, ultimately, the hound, all share common ground: They are all bound together as entities who "live and die as embodied beings"—a bond "that is the more powerful because it is 'unthinking' and in a fundamental sense unthinkable"—and as embodied beings, they are all subject to an "agential" reality.[13] "Life has become like that great Grimpen Mire", Watson realises (73): a "bog in which we are floundering" (89). Indeed, the Mire claims Stapleton's life, and nearly puts an end to Holmes'.

How, then, might we approach the novella in a way that enables us to take account, not only of the human–animal doublings it describes, but of the mutual and "mortal world-making entanglements" in which those relationships are grounded and through which they are transacted?[14] How, in turn, might we respond to the questions of agency and of representation that the novella raises? One answer is suggested by theories of new materialism, and, in particular, by the theories associated with Karen Barad and Donna Haraway. As Stacey Alaimo explains in her own reading of

Barad's "onto-epistemology", the "material turn [...] takes matter seriously", recognizing in both "human bodies and the environment" what Barad has called "a dynamic and shifting entanglement of relations".[15] Drawing on discoveries in the field of quantum physics, with its recognition of the co-constitutive inseparability of subject and object, Barad's theory of new materialism stresses the agentiality of the non-human (or more-than-human) world, and the way in which that world "intra-acts" with the human.[16] These intra-actions are so fundamental that, as both Haraway and Barad have argued, individual agencies do not precede, but in fact emerge from them.[17] Thus, identity emerges iteratively and performatively from these "phenomena"—these quantum entanglements—destabilizing "fixed and inherent" categories such as "'human' and 'nonhuman'".[18] From this perspective, there is only, as Haraway argues, a "subject- and object-shaping dance of encounters", each one generating new and specific versions of the self.[19]

Consequently, theories of new materialism stress the extent to which the human subject is imbricated in a world from which it is neither separate nor distinct, a decentring of the sovereign human subject that Doyle's novella itself enacts. Not only does the narrative build up a picture of the material, non-human world as alive and agential, underlining the extent to which that world influences characters such as Watson and even Holmes, but it also enacts a reciprocal shift in the status of the hound, as Doyle's great creation intra-acts with the novella's human characters. With their interlinked emphasis on the world's "radical aliveness" and on the quantum entanglements that in fact constitute and continually reconstruct identity, theories of new materialism therefore offer a useful means of reinterpreting a novella which itself breaks down the distinctions that distinguish the human from the non-human.[20] But as theories of new materialism also emphasize, "discursive practices" are themselves agential: They co-constitute the "material-social relations of the world".[21] This is not simply to point out that Conan Doyle's tales are material artefacts, or to draw attention to the way in which readers sometimes mistake Holmes for a real detective; in a further "entanglement of relations", Doyle's narrative materially affects the reader, and it does so in specific ways.[22] For example, the novella makes notable use "of Gothic imagery and rhetoric", which enables the narrative voice to suggest what a narrowly realist story might not: that "imagination [...] may exceed reason"; that character might find its mirror in "monstrous double[s] signifying duplicity and evil nature"; and that nature itself might be "alienating and full of menace".[23] All these features of the Gothic form a part of Doyle's story. Even that most characteristic of Gothic devices, the gloomy ancestral pile (here represented by Baskerville Hall) puts in an appearance. Similarly, the choice of a longer form for the story (in this case, a novella) enables the narrator to build suspense, whilst also disrupting the format with which fans of the Holmes short story were familiar: Although the tale opens in Baker Street, the action quickly moves from London to the wilds of Dartmoor; moreover, the great detective bows out of the narrative in chapter 6 (53), and does not reappear until the very end of chapter 11 (122).

As this brief discussion underlines, the suggestive power of Conan Doyle's hound depends on the way in which the novella first breaks down many of the distinctions that separate human from non-human—distinctions on which Holmes' success as a detective depends—whilst also disrupting the narrative framework within which Doyle's stories about Holmes are usually situated. As this chapter emphasises, that process of creative disruption is not simply a function of the interplay between Holmes, hound, and the other human characters in the novella, nor even of their shared immersion in the agential reality of Dartmoor; they emerge from the reader's own entanglement with the novella as a material-discursive practice.

Throughout the nineteenth century, Harriet Ritvo argues, "people systematically appropriated power they had previously attributed to animals, and animals became significant primarily as the objects of human manipulation".[24] That process of appropriation and exploitation was, however, accompanied by a kind of compensatory sentimentalisation of the animal, most apparent in the nineteenth-century adoption of the pet.[25] Increasingly, "animals—in particular dogs—came to be installed at the heart of the [...] bourgeois household".[26] "[T]he dog", argues Claire McKechnie, "had a special place in the hearts and minds of Victorians".[27] Unsurprisingly, many stories of the turn-of-the-century period allude to or include dogs, "especially those that fall within the detective genre".[28] The Sherlock Holmes stories are no exception: Dogs "unlock many secrets in Conan Doyle's mysteries".[29] In the Sherlock Holmes stories, they include a wolfhound, a spaniel, an Airedale, and Doyle's own lurcher-spaniel cross, who appears as Toby in *The Sign of the Four*; most are friendly and devoted.[30] Perhaps surprisingly, given his reputation as a "science-oriented" detective and "cold logician", Holmes is himself frequently compared to a dog.[31] Not only does he have "a special affinity with dogs", but is often presented as "the embodiment of a sleuthhound", seeking out a scent.[32]

According to Emma Mason, these canine references point to the reality of Holmes' "dog-like character, companionable and committed", but also "commonsensical and compassionate".[33] As Mason argues, Holmes approaches every case as if he were himself a dog, "frequently exonerates or celebrates the dogs he encounters", and has a "canine relationship" with the "houndishly loyal" Watson.[34] Drawing on Donna Haraway's "companion species theory", which explores "human relations with the non-human in order to make an argument for the contingency of subjectivity", Mason demonstrates that, far from being aloof from the everyday, material world, Holmes is, in fact, immersed in it, an "attachment to everyday life [that] is itself a marker of his dog-like character".[35] Paradoxically, it is precisely this attentiveness that distinguishes Holmes from (say) the often baffled Watson, and distances him from society in general. Yet it is also "the 'otherness' of Holmes' [...] personality" that binds him to Watson; each constitutively complements the other, and together, they "form a kind of companion species".[36] Otherness becomes, in Haraway's formulation, significant, and significant in all its particularity, its "*specific difference*".[37]

Although Mason does not include it in her analysis, Haraway's later work is also relevant to an understanding of Doyle's novella. In *When Species Meet* (2008), Haraway extends her interest in "co-constitution [and] contingency" to the "mortal world making entanglements" from which identities emerge.[38] As Haraway notes, it is "patterns of relationality and, in Karen Barad's terms, intra-actions [...] that need rethinking".[39] Beyond terms such as human (or transhuman or posthuman) and animal—each just another "troubled category"—lies that "dance of encounters" from which identity emerges, creating in the moment of intra-action a haunting sense of "doubleness".[40] Being, she argues, is caught up in these entanglements; it derives from these phenomena, these "contact zones".[41]

The relevance of Haraway's theorization of contact zones to Doyle's story is underlined by the second aspect of her argument. "What Bruno Latour calls the Great Divides between what counts as nature and as society, as nonhuman and as human" are, suggests Haraway, aspects of a single divide, which opposes "all Others" ("the colonized, the enslaved, the noncitizen, and the animal") "to rational man".[42] Stapleton is a case in point. A "rational man", he is master not only of the hound, but also of his beaten, subjugated wife (154), both of whom he has enslaved; as the case unfolds, however, Stapleton is himself repositioned as "a creature"

(126); it is the monstrous Stapleton who is "the real hound of the Baskervilles".[43] The escaped convict, Selden, is a still more extreme instance of this regression from the dominant norm of "rational man".[44] With his "sunken animal eyes" (131) and his "terrible animal face" (97), the narrative depicts him as a brute, and consequently "brutal" (94). By contrast, "rational man" finds his (perhaps supreme) expression in Holmes himself, "cold, incisive, ironical" (122). It is part of the novella's power, however, that it transgresses this paradigm, in part by suggesting that Holmes is powerfully affected by his immersion in the "contact zone" of Dartmoor. As Watson remarks, on the night that the escaped convict Selden is (literally) hounded to his death, "I knew from the thrill of his [Holmes'] voice that he, the man of iron, was shaken to the soul" (128).

In Doyle's tale, therefore, rational man is "entangled" in an agential reality that undermines his supremacy over it; he becomes simply another of many "ordinary beings-in-encounter".[45] As if to highlight that entanglement, there are frequent references to nets in the novella. Chapter 13 is entitled "Fixing the Nets", and Holmes refers to his own net, "closing upon him [Stapleton], even as his are upon Sir Henry [Baskerville]" (127), Stapleton's intended victim. It is Watson, however, who identifies the "net'—or "unseen force" (121)—in which they are all caught up.

For Watson, however, it is no easy matter to describe what he experiences. Nearing Dartmoor, Watson spies "a grey, melancholy hill, with a strange jagged summit, dim and vague in the distance, like some fantastic dream" (54). The hill marks his first glimpse of the moors, but his description of it as strange, fantastic, jagged yet vague also marks the difficulty he will have in reconciling his conventional notion of the more-than-human world as "mute and immutable" with his embodied experience of it.[46] To Watson, the moors initially resemble a "desert" (119), "desolate" (56) and devoid of life (119). Lacking the lush fertility of the "peaceful and sunlit country-side" which he is accustomed to calling "Nature" (55), their desolation is itself suggestive: because empty, they appear "lonely" (115) and "melancholy" (54, 55, 60, 72, 105). To call the moors melancholy is, however, to fall into the trap of the pathetic fallacy: Watson has attributed a human emotion to an inanimate object. But as Watson realises, the moors are not, in fact, "a passive surface awaiting the mark of culture".[47] As his reports to Holmes make plain, the moors are part of a world that is vigorously alive. The narrative describes trees that "moaned and swung" (60) in the wind, "twisted and bent by the fury of years of storm" (56); it refers to the wind "moaning" (94) and "whistling," rain "rustling on the ivy" (105), and squalls "trailing [...] grey wreaths" (105) over "craggy hills" that "loom" out of the darkness (96); the hills are themselves like "long green rollers, with crests of jagged granite foaming up unto fantastic surges" (66). The "rolling" (71), "swelling plain" (68) is to Watson an "active factor".[48] It is alive; Watson even refers to its "russet face" (105).

Like the "spectral hound" itself, this "*dynamic articulation/ configuration of the world*" seems to lie "outside the ordinary laws of Nature" (100).[49] More accurately, it exceeds Watson's conventional concept of "Nature" as a "fixed substance" or "fixed essence".[50] To the contrary, the moors are an instance of "substance in its intra-active becoming", and through his own intra-actions with it, Watson is materially affected, physically, mentally, and emotionally.[51] For Watson, the agentiality of this "enormous wilderness of peat" (98) is unsettling, and for Watson in his guise as narrator, problematic. As Alaimo notes, agency is usually linked to "rational—and thus exclusively human—deliberation".[52] To invoke agency is to suggest a teleology: it implies that "nature" has a purpose. As Barad has insisted, however, matter is "*not a thing but a doing*:" its "purpose" is not defined by any particular end state; rather, its purpose lies in its "*ongoing materialization*".[53] Caught up in his own experience of the moors, and troubled by his impression of them as "wild", "bleak" (105), "harsh" (56), but also "inscrutable" (99), Watson cannot avoid attributing to the moors some kind of purpose, intent, or motive; nor can he avoid the inference that, because the moors are inhuman, their intent is necessarily antihuman. Whilst Holmes has asked him "simply to report facts", and not "bias [his] mind by suggesting theories or suspicions" (51), Watson finds it impossible to describe the moors without also conveying his sense that they are "sinister" (55) and "ill-omened" (148). "[C]onscious of shadows all around me" (73), and with "all traces of modern England" left behind (75), Watson feels "the spirit of the moor sink into [his] soul" (75).

Thus, and as Watson journeys deeper into the moors, his description of the "desolate plain" (56) multiplies phrases or words that underline a feeling of its strangeness and severity, but also a growing feeling that it is charged with menace; the fact that the escaped murderer, Selden, is now hiding out on the moors, is, for Watson, "enough to complete the grim suggestiveness of the barren waste, the chilling wind, and the darkling sky" (56). That penultimate term ("darkling") borrows from Matthew Arnold's mid-Victorian poem, "Dover Beach," in which, very famously, a "darkling plain" is "[s]wept with confused alarms of struggle and flight" as the "Sea of Faith" recedes. To Watson, the moors are no less "God-forsaken" (75), neatly recalling his own, opening discussions with Holmes, in which both entertain the possibility that the "hell-hound" is simply a material manifestation of "the Father of Evil" (23); and whilst God's absence need not mean Satan's presence, as Richard Davenport-Hines points out, the Gothic "easily [fills] the imaginative vacancy".[54]

By the time that Watson and his travelling companion finally reach Baskerville Hall, itself a place of "gothic spookiness", Watson's narrative has therefore taken on the tone and many of the tropes associated with the Gothic.[55] As John Scaggs points out, mystery and detective fiction is "closely allied" to the Gothic; indeed, the "modern crime thriller can be traced from the Gothic novel".[56] In particular, there is a strong parallel between the Gothic, and crime narratives in which "the past threatens the social order in the present".[57] With its backstory of the centuries-old legend of Hugo Baskerville, punished for his wickedness by a "demon dog" (160), Doyle's novella follows this pattern. But as Davenport-Hines adds, the Gothic also denotes "terror, mystery, despair, malignity, human puniness and isolation".[58] Moreover, and in opposition to the Enlightened faith in humankind's "power over nature", the Gothic reasserts "the power of natural forces over man", forces that increasingly form a part of Watson's own "dark imaginings" (105) as he walks "upon the sodden moor" (105).[59] The moors may seem "like some fantastic dream" (54), but as the ever-practical Watson insists, the impression they give is "not entirely a question of imagination" (61). Nor is Watson alone in feeling immersed in (and entangled with) a "jagged and sinister" (55), but also agential reality. As Watson discovers, Holmes is abroad on the moor, and he too has felt a presence at work there that upsets his plans. It also clouds his judgement. Having already risked Sir Henry's life by continuing to expose him to a threat that he knows to be real—and the reality of that risk is underlined when, mistaking Selden for Sir Henry, the hound chases the convict to his death—Holmes nevertheless insists on using Sir Henry as bait. This time, his plan is thrown into disarray by the fog that drifts in from Grimpen Mire (148–149), a fog that is itself agential, as Watson's description of it emphasises: "moving", "curling", and "crawling" (149), it sweeps "slowly and inexorably on" (150), allowing the hound to catch Sir Henry. It is pure luck that the baronet survives the encounter. What Dr Mortimer, himself a "man of science" (17), calls "the settled order of Nature" (22), is nowhere to be found; instead, "the world's radical aliveness" has come "to light".[60] Indeed, Holmes shares in Watson's Gothicized sense of "some monstrous villainy, half-seen, half-guessed", looming through the darkness (126). Holmes "maintains a Voltairean scepticism about the forces of evil whilst he is London", W. W. Robson observes, but once out on the moors, "the sarcastic note drops out of Holmes' conversation and does not return".[61]

As Julian Symons notes, the success of *The Hound of the Baskervilles* turns on the way in which Doyle "makes us feel the terror and loneliness of the Devon moors".[62] Whilst Gothic tropes make an important contribution to that effect, however, they also act to limit and constrain the way in which the narrative constructs meaning. According to Davenport-Hines, the Gothic emerges from "a historical continuum reflecting irrationalism [and] pessimism"; it accords with a shared impression that an agential world is necessarily *un*natural and hence sinister, even malicious.[63] But this negative construction also extends to the hound, the "fiend dog" (64) of legend. As McKechnie argues, Doyle's hound participates in a wider "Gothicization of the canine in literature", driven not only by general concerns about degeneration, noted earlier, but also by a growing public fear of rabies.[64] Although the language of the Gothic may, as McKecknie notes, be ambiguous, and whilst the Gothic may serve to disrupt the "bodily limits between human and animal", Watson's interpretation of the Gothic mode emphasises the sinister, the malevolent, and the threatening.[65] A hound may be defined simply as a hunting dog, but in the novella, this "huge black creature" (150) is, perforce, "a brute", a "beast" (13), "black, silent, and monstrous" (78), "dark" of form, "savage" of face (151), with eyes "deep-set, cruel" (152). But whilst Watson describes its "strange cry" as "strident, wild, and menacing" (95), he also describes that sound as a "sad moan" (95), a description that mirrors his first impression of its howl, which he describes as "[a] long, low moan, indescribably sad" (68). It is as if the dog-like Watson has sensed something of the hound's own, embodied reality—its own predicament—chained up in an abandoned mine in the middle of the Mire. As Watson and Holmes later discover, the "starved and mistreated Baskerville hound" is still more fundamentally a product of human interference, and of its subjection to the human.[66] The hound is, in fact, the crossbred (152) corollary of "a wider transformation" in the status of the Victorian dog, as dog fanciers "celebrated the power to manipulate the raw material of breeding dogs and manufacture something novel".[67] The hound emerges as a kind of Frankenstein, but like Frankenstein itself, the hound is not (or not quite) the "manufactured monster" that the Gothic demands, and its heightened vocabulary of evil menace assumes.[68] To the contrary, and as O'Gorman notes, "readers will deplore the treatment of the hound".[69]

The limitations of Watson's Gothicized narrative reflect what Wolfe describes as a subjection to language: language constitutes a form of finitude, "a radically ahuman technicity" that limits the ways in which we think.[70] It is something the human has in common with the animal, a form of constraint embodied in language, or reflected in the lack of it.[71] There is, however, another form of finitude that challenges the human sense of separateness and supremacy. Out walking the morning after his arrival at Baskerville Hall, Watson hears the cries of a moor pony as it falls into the nearby Grimpen Mire, a "treacherous" bog (69) from which there is no escape (67) ("[i]t turned me cold with horror') (67). Later, Holmes and Watson find themselves fighting the Mire's "tenacious grip" (155): "it was if some malignant hand was tugging us down into those obscene depths, so grim and purposeful was the clutch in which it held us" (156). The reckless Holmes is particularly lucky to escape with his life: spying a piece of evidence, "Holmes sank to his waist as he stepped from the path to seize it, and had we not been there to drag him out he could never have set his foot upon firm land again" (156).

The Mire is a reminder that, in the world of the novella, human and non-human alike are subject to the same "material phenomena", and both share the same mortal weaknesses.[72] In common with the non-human "other", Wolfe argues, we experience a "physical vulnerability, embodiment, and eventually mortality" that itself constitutes a form of finitude linking human and animal; in the case of Doyle's novella, that bond also binds Watson, Holmes and the hound.[73] It follows from this strong sense of mutual vulnerability that the hound is not supernatural. Holmes insists from the outset that the hound is entangled with the same world that links him to Watson: The "Baskerville demon" might well be "luminous, ghastly, and spectral" (22), as witnesses to it claim, but the footprint it leaves is "material" (23, 100), just as its ancestral predecessor "was material enough to tug a man's throat out" (23). In turn, its materiality points to its mortality. When, during their fateful encounter with the hound, Holmes and Watson shoot at and wound it, its "cry of pain" tells them that it is indeed vulnerable: and "[i]f he was vulnerable he was mortal, and if we could wound him we could kill him" (151).

By the time the narrative reaches its conclusion, therefore, it has repeatedly insisted on the frailty and vulnerability of its subjects: on the weaknesses in Holmes' abstract theorizing, which fails to take full account of "matter's dynamism", and in so doing exposes his client to real and unnecessary risk; on his own vulnerability as a being who cannot in the end absent himself from his own body, or ignore the vulnerabilities that make it prey to the haunting presence of the hound; and finally, on the vulnerability of the hound itself.[74] Significantly, the capacity to feel but above all to suffer establishes a connection between human and animal that blurs the ethical boundary separating the two, as philosophers from Jeremy Bentham to Peter Singer and Jacques Derrida have insisted.[75] Yet it also becomes apparent that, just as humans have manipulated the behaviour of the hound, so Holmes has himself been materially altered and affected. "Bodies are not objects with inherent boundaries and properties", notes Barad.[76] Identities emerge iteratively, through "the often unpredictable and unwanted [intra-actions] of human bodies, nonhuman creatures, ecological systems, […] and other actors".[77] For Holmes, argues Neil Pemberton, the effect is dramatic: "[o]n the moors", Pemberton suggests, "Holmes takes on the characteristics of the abominable hound", "blurring the lines between detective and criminal, human and animal".[78] "'[K]ennelled' [in] a stone hut", Holmes "hides secretly […] hunting the culprit in the nocturnal hours".[79] He is "like a bloodhound, independent and distanced from his own master, Sir Henry"; "as a result, the detective himself is mistaken for a spectral being or a throwback".[80] In the "contact zone" of the moor, the "creature of reason" has emerged as "a more complex and ambiguous" entity, "associated with hunting, instinctual knowledge, and animal behaviour".[81]

Humans, Alaimo contends, are "permeable, emergent beings, reliant upon the others within and outside our porous borders".[82] As Pemberton emphasises, the novella's blurring of lines suggests awkward questions about the animal qualities of Holmes, whom Watson has elsewhere described as a "sleuth-hound," even as a "pure-blooded […] foxhound".[83] Increasingly, Holmes resembles a "hybrid".[84] In other words, Holmes' identity is itself changing, developing; as Barad has argued, it is a function of a constant, shifting pattern of relations, and our awareness of that relationality is limited only by "humanity's own captivity within language".[85] Yet the hound's emergent identity remains problematic, perhaps because itself obscured by the representational limitations—that "captivity within language"—to which Barad and also Wolfe refer. W. W. Robson argues that the hound is anomalous; it confounds the reader's expectations by inverting the conventional, comforting reading of the dog as human helpmate.[86] In fact, the contrary argument might be made that as Stapleton's "savage ally" (156) the hound confirms rather than confounds the history of "companion species".[87] In spite of the terror it strikes into Holmes, Watson, and others, the hound's existence is caught up with Stapleton's, to whom it remains loyal: At his bidding, it hounds a convict to his death (128–131), and nearly kills Sir Henry (151); it risks, and in the end loses its own life, for the sake of its human keeper. But the still more unsettling possibility is that the hound has neither done Stapleton's bidding only because coerced into doing so, nor, as the devoted canine of cliché, surrendered its own agency out of loyalty to him. The novella's final chapter, "A Retrospection", briefly sketches what Holmes and Watson can ascertain of the mechanics of the relationship between Stapleton and the dog, including the dog's purchase and kennelling in the Mire (160), and the way it was trained by Stapleton to recognise Sir Henry's scent, and "half-starved" (167) to encourage its savagery. But is its human handling sufficient to account for that savagery? "A Retrospection" suggests otherwise. Even when first bought from the dealers in London, Watson notes, the dog was "the strongest and most savage in their possession" (160). Perhaps, the narrative implies, there is no explanation for the animal's behaviour, or none that is dependent on human agency.[88]

Consequently, Doyle's depiction of the hound offers a disturbing challenge to the way in which turn-of-the century British society increasingly regarded "the first and most intimate companion of man", as the dog was described by Nathaniel Shaler in 1895.[89] The narrative opens up the question of what, if anything, humans have ever truly known about dogs—or each other—even in the midst of their intra-action. The savagery of the hound is, like Stapleton's "monstrous villainy" (126), unaccountable. The counterpoint to the novella's ingenious blurring of human/animal distinctions, and its assertion of similarities and parallels, is the strong sense given by Watson's narrative that, in spite of what we think we know about dog or animal, each is in its specific particularity unknowable, each an "unsubstitutable singularity".[90] Arguably, this is the real source of the fears felt by Doyle's characters: Dogs may have been domesticated, but is this anything more than a temporary arrangement, subject to the dog's own continued compliance? The "smouldering glare" (150) of the hound reasserts a fact that its human adversaries had forgotten but the narrative reasserts: that the animal is sufficiently itself to have its own "point of view" as it regards us.[91] Therein lies its own radically alien and potentially threatening nature.

In "And Say the Animal Responded?" Jacques Derrida spoke of "the whole anthropomorphic reinstitution of the superiority of the human order over the animal order, of the law over the living".[92] Certainly Holmes (as the idiosyncratic embodiment of that law) seems to symbolise the reassertion of human distinction and dominance. But is that process of reassertion entirely successful? With its intricate and powerful portrayal of the links between people and place, human and animal, *The Hound of the Baskervilles* invokes what Haraway calls "layers of history, layers of biology, layers of naturecultures".[93] In the quantum entanglements that Doyle's novella enacts, hound and human protagonists are performatively revealed as bodies "in which social power and material/ geographic agencies intra-act".[94] United by that shared field of experience, humans are glimpsed becoming animal-like, and animals, like humans; both humans and animals have experienced a shared vulnerability and even mortality, but also a mutual subjection to the limitations of language.[95] These similarities and commonalties are underlined by the extent to which all the principal protagonists in the novella are affected by the agential reality of the moors, the contact zone in which their "world-making entanglements" have taken place.[96] Here, "diverse bodies and meanings coshape one another".[97] Ironically, therefore, the tale that Doyle so cleverly unfolded subverts its own premise: it collapses, rather than reasserting, "the Great Divides" to which Latour refers.[98] Nevertheless, the hound that emerges from its encounters represents a kind of material-discursive surplus, inexplicable except in its own terms, without human reference; it exists beyond the reach of Holmes' reductive methodology, and even in death, it remains a haunting presence. The beast may be dead, but the world of which it formed a part still exists, as both literal, physical place (the moors) and as haunting, psychological space (the atavistic fears which the novella invokes but never finally puts to rest). In short, what makes *The Hound of the Baskervilles* one of the great detective stories is not just the way in which all the main characters are profoundly entangled with and affected by the hound, but the way in which, ultimately, the hound (re)asserts its own, bewilderingly strange, entirely unknowable self, a self that death itself cannot silence.

Notes

1. Doyle, *Hound*, 150–151. All other references to the novella will be found in the text.
2. Mason, "Dogs, detectives," 293.
3. O'Gorman, "Introduction," 17.
4. Howell, *At Home*, 3.

5. Clausen, "Sherlock Holmes," 83; McKechnie, "Man's Best," 117.
6. Jones, *The Phantom*, 7, 20.
7. McKechnie, "Man's Best," 131; Clausen, "Sherlock Holmes," 83.
8. Clausen, "Sherlock Holmes," 83.
9. Wolfe, *Posthumanism*, 99; Pemberton, "Hounding Holmes," 464.
10. Wolfe, *Posthumanism*, 127, xv; O'Gorman, "Introduction," 25.
11. Pemberton, "Hounding Holmes," 465.
12. O'Gorman, "Introduction," 22.
13. Wolfe, *Posthumanism*, 123; Barad, *Meeting*, 32.
14. Haraway, *When*, 4.
15. Alaimo, *Bodily*, 21, 6, 3; Barad, *Meeting*, 5.
16. Barad, *Meeting*, 32.
17. Haraway, *When*, 4–5; Barad, *Meeting*, 33. Drawing on the work of philosopher-physicist Nils Bohr, Barad derives a theory of agential realism that highlights the importance of phenomena—not independent things or entities—as the "basic units of reality"; phenomena mark the "ontological inseparability of intra-acting components" through which discrete, separable agencies or entities emerge, and the "neologism 'intra-action'" is used to underline this productive interplay; Barad, *Meeting*, 33. Haraway's concept of "contact zones" (see hereafter) overlaps with Barad's definition of phenomena.
18. Barad, *Meeting*, 33, 32.
19. Haraway, *When*, 4.
20. Barad, *Meeting*, 33.
21. Barad, *Meeting*, 34, 35.
22. Barad, *Meeting*, 35.
23. McKechnie, "Man's Best," 131; Botting, *Gothic*, 3, 2, 2.
24. Ritvo, *The Animal*, 2.
25. Ritvo, *The Animal*, 2–4; Howell, *At Home*, 176.
26. Howell, *At Home*, 11.
27. McKechnie, "Man's Best," 118.
28. Mason, "Dogs, detectives," 290.
29. Mason, "Dogs, detectives," 290, 291; see also Pemberton, "Hounding Holmes," 461–2.
30. Robson, "Introduction," xxiii.
31. O'Brien, *Scientific Sherlock*, ix; Mason, "Dogs, detectives," 291, 290.
32. Pemberton, "Hounding Holmes," 455.
33. Mason, "Dogs, detectives," 291.
34. Mason, "Dogs, detectives," 291.
35. Mason, "Dogs, detectives," 291.
36. Pemberton, "Hounding Holmes," 455; Mason, "Dogs, detectives," 295.
37. Haraway, *Companion*, 7, 3 (emphasis in the original).

38. Haraway, *Companion*, 7; Haraway, *When*, 4.
39. Haraway, *When*, 17.
40. Haraway, *When*, 17, 4.
41. Haraway, *When*, 4.
42. Haraway, *When*, 9, 18.
43. O'Gorman, "Introduction," 25.
44. Haraway, *When*, 18.
45. Haraway, *When*, 5.
46. Barad, *Meeting*, 183.
47. Barad, *Meeting*, 183.
48. Barad, *Meeting*, 183.
49. Barad, *Meeting*, 151 (emphasis in the original).
50. Barad, *Meeting*, 151, 183.
51. Barad, *Meeting*, 183.
52. Alaimo, *Bodily*, 143.
53. Barad, *Meeting*, 151 (emphases in the original).
54. Davenport-Hines, *Gothic*, 1.
55. O'Gorman, "Introduction," 17.
56. Scaggs, *Crime*, 35, 106.
57. Scaggs, *Crime*, 16.
58. Davenport-Hines, *Gothic*, 1.
59. Davenport-Hines, *Gothic*, 3.
60. Barad, *Meeting*, 33.
61. Robson, "Introduction," xvii.
62. Symons, *Bloody*, 77.
63. Davenport-Hines, *Gothic*, 3.
64. McKechnie, "Man's Best," 116, 115.
65. McKechnie, "Man's Best," 117.
66. Mason, "Dogs, detectives," 298.
67. Pemberton, "Hounding Holmes," 454.
68. McKechnie, "Man's Best," 131.
69. O'Gorman, "Introduction," 22.
70. Wolfe, *Posthumanism*, 73, 119.
71. Wolfe, *Posthumanism*, 135.
72. Barad, *Meeting*, 34.
73. Wolfe, *Posthumanism*, 118.
74. Barad, *Meeting*, 151.
75. Derrida, *Animal*, 27.
76. Barad, *Meeting*, 153.
77. Alaimo, *Bodily*, 2.
78. Pemberton, "Hounding Holmes," 465.
79. Pemberton, "Hounding Holmes," 465.

80. Pemberton, "Hounding Holmes," 465.
81. Haraway, *When*, 4; Pemberton, "Hounding Holmes," 466.
82. Alaimo, *Bodily*, 156.
83. Doyle, *The Complete*, 185, 31.
84. Pemberton, "Hounding Holmes," 465.
85. Barad, *Meeting*, 137.
86. Robson, "Introduction," xxiii.
87. Haraway, *Companion*, 5.
88. My own point is somewhat different to McKechnie's argument that "the dog has been designed by humans to attack and destroy at will—*bred* to kill" ("Man's Best," 131), and that it is "the 'strongest and most savage' the breeders could produce" (131). Although we might speculate that the hound's savagery is the result of an experiment in cross-breeding, the narrative supplies no evidence that the animal was deliberately bred as "a biological weapon" (McKechnie, "Man's Best," 131).
89. Qtd, Howell, *At Home*, 10.
90. Derrida, *Animal*, 9.
91. Derrida, *Animal*, 11.
92. Derrida, *Animal*, 136.
93. Haraway, *Companion*, 2.
94. Alaimo, *Bodily*, 63.
95. Wolfe, *Posthumanism*, 118, 73.
96. Haraway, *When*, 4.
97. Haraway, *When*, 4.
98. Haraway, *When*, 9.

Works Cited

Alaimo, Stacy. *Bodily Natures: Science, Environment, and the Material Self.* Bloomington, IN: Indiana University Press, 2010.

Barad, Karen. *Meeting the Universe Halfway.* London: Duke University Press, 2007.

Botting, Fred. *Gothic.* London: Routledge, 1996.

Clausen, Christopher. "Sherlock Holmes, Order, and the Late-Victorian Mind." In *Critical Essays on Sir Arthur Conan Doyle*, edited by Harold Orel, 66–91. New York, NY: G. K. Hall & Co, 1992.

Davenport-Hines, Richard. *Gothic: 400 Years of Excess, Horror, Evil and Ruin.* London: Fourth Estate, 1998.

Derrida, Jacques. *The Animal That Therefore I Am*, edited by Marie-Louise Mallet, translated by David Wills. New York, NY: Fordham University Press, 2008.

Doyle, Sir Arthur Conan. *The Complete Sherlock Holmes.* London: Penguin, 2009.

Doyle, Arthur Conan. *The Hound of the Baskervilles*, edited by W. W. Robson. Oxford: Oxford University Press, 2008.

Haraway, Donna. *Companion Species Manifesto: Dogs, People, and Significant Otherness*. Chicago: Prickly Paradigm Press, 2003.
Haraway, Donna J. *When Species Meet*. London: University of Minnesota Press, 2008.
Howell, Philip. *At Home and Astray: The Domestic Dog in Victorian Britain*. London: University of Virginia Press, 2015.
Jones, Kelvin I. *The Phantom Hound*. Norfolk: Oakmagic, 2006.
Mason, Emma. "Dogs, detectives and the famous Sherlock Holmes." *International Journal of Cultural Studies* 11, no. 3 (2008): 289–300.
McKechnie, Claire Charlotte. "Man's Best Fiend: Evolution, Rabies, and the Gothic Dog." *Nineteenth Century Prose* 40, no. 1 (Spring 2013): 115–140.
O'Brien, James. *The Scientific Sherlock Holmes: Cracking the Case with Science & Forensics*. Oxford: Oxford University Press, 2013.
O'Gorman, Francis. "Introduction." In Arthur Conan Doyle, *The Hound of the Baskervilles with "The Adventure of the Speckled Band,"* edited by Francis O'Gorman, 13–42. Plymouth: Broadview, 2006.
Pemberton, Neil. "Hounding Holmes: Arthur Conan Doyle, Bloodhounds and Sleuthing in the Late-Victorian Imagination." *Journal of Victorian Culture* 17, no. 4 (2012): 454–467.
Ritvo, Harriet. *The Animal Estate: The English and Other Creatures in the Victorian Age*. Cambridge, MA: Harvard University Press, 1987.
Robson, W. W. "Introduction." In Arthur Conan Doyle, *The Hound of the Baskervilles*, edited by W. W. Robson, xi–xxix. Oxford: Oxford University Press, 2008.
Scaggs, John. *Crime Fiction*. Abingdon: Routledge, 2005.
Symons, Julian. *Bloody Murder*. Harmondsworth: Penguin, 1975.
Wolfe, Cary. *What is Posthumanism?* London: University of Minnesota Press, 2010.

Wolverines, Werewolves and Demon Dogs: Animality, Criminality and Classification in James Ellroy's *L.A. Quartet*

Nathan Ashman

James Ellroy is an eccentric and divisive popular novelist. Since the publication of his first novel *Brown's Requiem* in 1981, Ellroy's outré "Demon Dog" persona and his highly stylised, often pornographically violent crime novels have continued to polarise both public and academic opinion. Frequently profane and unbridled by political correctness, Ellroy's public appearances are regularly punctuated by a barrage of racial invective, casual homophobia and spouts of wild dog barking, creating a profound tension between his brash and unpredictable public identity and the considered meticulousness of his art. Indeed, Ellroy typically opens his readings with the following refrain: "Good evening, peepers, prowlers, pederasts, panty sniffers, punks and pimps. I'm James Ellroy, the demon dog, the foul owl with the death growl, the white knight of the far right and the slick trick with the donkey dick".[1] It is perhaps no surprise, then, that Ellroy's

N. Ashman (✉)
University of East Anglia, Norwich, UK
e-mail: N.Ashman@uea.ac.uk

R. Hawthorn, J. Miller (eds.), *Animals in Detective Fiction*, Palgrave Studies in Animals and Literature,
https://doi.org/10.1007/978-3-031-09241-1_4

"Demon Dog" persona has tended to overshadow the reception and appraisal of his work, with many critics perceiving the stark presentations of race, sexuality and gender in Ellroy's texts as a straight reiteration of the writer's own skewed far-right agenda. Whilst Ellroy's personal politics remain a point of contention, such stark articulations of "animalistic" behaviour are a central facet of Ellroy's self-styled and hyperbolised public identity and are symptomatic of a broader preoccupation with animality, criminality and sexual/racial classification in his work.

Representations of beastliness and monstrosity—both literal and figurative—are a recurrent feature of Ellroy's work and represent the culmination of a long-standing fascination with animals that Ellroy himself traces back to his early obsession with "animal stories".[2] Initially providing an outlet for what he describes as a "wrenchingly tender and near-obsessive" love of animals, Ellroy eventually boycotted these narratives due to their propensity to depict suffering, cruelty and death.[3] The extent to which this has influenced Ellroy's subsequent, often warped, depiction of animalism in his fiction is contestable. Although Ellroy's long-standing tenderness towards animals is well documented, his texts are punctuated with various instances of animal cruelty and bestiality, acts which become symptomatic of a more pervasive culture of monstrous violence, sexual savagery and atavistic desires. Such ambivalence seems to be symptomatic of Ellroy's relationship with, and codification of, animals more broadly. Indeed, when questioned about the instances of animal mutilation in his texts by the critic Steven Powell, Ellroy replied: "I dig animals. I also dig a good steak. I love animals. I wouldn't go shoot one, but I'll sure as shit eat the piece of steak you put on my plate".[4]

This sense of contradiction is no more evident than in Ellroy's *L.A. Quartet*, a four-volume criminal and political history set against the backdrop of post-war Los Angeles's urban expansion.[5] In these texts, Ellroy depicts a vivid tapestry of literal and symbolic animality. From the savage brutality of "Wolverine Killer" Coleman Healy in *The Big Nowhere* (1988) to the warped, ornithological mutilations of child murderer Loren Atherton in *L.A. Confidential* (1990), Ellroy's novels trace a potent convergence between animality and deviancy. This convergence becomes forcefully energised by rigid racial, sexual and political classifications, as manifestations of animality are repeatedly articulated via stark depictions of homosexuality, communism and "blackness". Depictions of race and homosexuality in particular are underlined by an essentialist rhetoric that aligns the predilection for violence and deviancy with certain predisposed and atavistic traits, both physiognomic and genetic. In both cases, the language of animality and classification becomes intimately bound up with "images of corruption and infection".[6]

Of course, this conjunction between animality and human discourses of otherness is by no means a new strand in the writing and analysis of detective texts, or indeed, literary criticism more broadly. Since the brutal murder of Madame L'Espanaye and her daughter at the hands of a vicious escaped orangutan in Poe's "The Murders in the Rue Morgue", fictional depictions of criminality have been recurrently energised by shifting anxieties regarding animality, identity and typology. As Christopher Pittard suggests, the "identification of criminality with physiology" was the basis of "late nineteenth century criminology" and operated to reduce the act of detection to "one of diagnosis".[7] Pittard refers specifically to the influence of criminal anthropologist Cesare Lombroso, whose work on the association between physiognomy and criminality harnessed broader cultural anxieties regarding questions of degeneration in the wake of emerging theories on evolution. Ronald Thomas sees these contexts as firmly rooted in Arthur Conan Doyle's *The Sign of Four*, arguing that Holmes is able to solve the case by "recognizing the culprits foreignness in the traces of the criminal body left at the scene of the crime".[8] For Thomas, this is indicative of recurrent patterns of association between criminality and racial inscription in the Holmes narratives and late nineteenth-century crime fiction more broadly.

Historically then, these codifications of criminality in detective fiction have become implicitly inscribed by discourses of race and degeneracy. This is no truer than in the hard-boiled American crime novel, where "other" races and "deviant sexualities" are frequently contained and delimited by monolithic mechanisms of white male power. Often centred around a brutal contestation of urban space that sees "otherness" pushed to the margins, Liam Kennedy argues that the hard-boiled genre has "responded to a 'dark Africanist presence' by adopting a parasitic relationship to blackness".[9] This does not apply exclusively to depictions of race; the hard-boiled novel has a long tradition of producing comparable engenderings of homosexuality, with a pointed emphasis on questions of masculinity and authenticity. One need only look at Chandler's depiction of Carroll Lundgren in *The Big Sleep* to gauge the manoeuvrings of white heterosexual power underpinning such narratives. After Marlowe is punched by Lundgren, he disdainfully remarks: "It was meant to be a hard punch, but a pansy has no iron in his bones, whatever he looks like".[10]

Through their parallel representations of deviancy and social classification, ostensibly Ellroy's texts can be seen to conform to a tradition of hard-boiled detective fiction that operates to reinscribe the dominance and authority of white heterosexual masculinity within hierarchical structures of social power. As Josh Meyer suggests, in Ellroy's work, depictions of criminality are "largely expressed through contesting articulations of blackness, Jewishness, gangsterism, homosexuality and other allegorical renditions of type".[11] Such accusations of a lack of variety and nuance to Ellroy's representation of black and/or "minority" characters have been levelled at his work on number of occasions and remain a consistent point of contention in much criticism regarding Ellroy's work. Yet, although Ellroy's use of derogatory and often uncomfortable stereotypes would, on the surface, seem to reinscribe these disparities of power long established in the genre, this essay will argue that Ellroy's depictions of atavism, criminality and deviancy are in fact more complex than they first appear. Through its exposure of the brutality and corruption at the centre of white characters and white institutions, the *L.A. Quartet* ultimately operates to subvert the logic that underpins such oppressive discursive practices, destabilising the paradigm that positions whiteness in opposition to the violence, animality and degeneracy of the atavistic "other".

Whilst this will comprise one strand of my analysis of animality in this essay, it still only goes some way towards understanding Ellroy's preoccupation with, and representation of, literal and symbolic animals. Indeed, despite Ellroy's subversion of the association between animality and "otherness" identified above, such an argument still risks compressing the vast complexity and variety of animal life into a generic category of "animality". The timely emergence of animal studies in recent years has recognised the importance of moving beyond what Derrida would term as the "asininity" of the concept of the animal as a singular category, interrogating instead the history and materiality of human–animal relationships to produce a new discourse that acknowledges animals as autonomous beings distinct from anthropogenic experience.[12] For Paul Waldau, one of the greatest opportunities derives from outlining and confronting the "powerful but dysfunctional form of human centeredness" or "human exceptionalism" that continues to dominate "much thinking in certain circles".[13] The pervasiveness of such an attitude not only continues to prevent an examination of the skewed hierarchy that has defined human–animal relations in the past, but also overlooks the "obvious limitations we, as humans, have in grasping the features of some other animals' lives".[14]

In the study of literary animals, Phillip Armstrong equally emphasises the importance of moving "beyond reading animals as screens for the reflection of human interests and meaning", of transgressing this form of mirroring that has historically, he argues, been the dominant way of "treating cultural representations of animals".[15] Crucially, Armstrong argues that the analysis of literature must not attend merely to "what animals mean to humans, but to what they mean themselves; that is, to the ways in which animals might have significances, intentions and effects quite beyond the designs of human beings" (3). Of course, the possibility of producing a greater knowledge of "non-human" agency through the reading of texts necessitates a reconceptualisation of our very understanding of what agency means. As Armstrong suggests, the mere suggestion of a form of animal cognition distinct from "the human" immediately "invites allegations of anthropomorphism" (3). Yet, this very convergence between agency and rational thought—or, the delimitation of agency as the province of the human—is born out of a "humanist" enlightenment philosophy "within which these traits came to define the human as such" (3). Thus, our understanding of what anthropomorphism means is predicated on a humanist conceptualisation of agency. The need to understand agency outside of, or beyond, notions of rationality and consciousness is central to the construction of what Armstrong identifies as a "mode of analysis that does not reduce the animal to a blank screen for the projection of human meaning" (3).

The second half of this essay will therefore endeavour to analyse the literal representation of animals in Ellroy's work, with a particular focus on his depiction of wolverines in *The Big Nowhere*. Whilst the political dynamics of animality as a generic symbol of otherness and deviancy are, on the whole, further solidified by the role actual animals play in Ellroy's work, the text also establishes a parity between the entrapped and exploited wolverines and the similarly disfranchised and abused serial killer Coleman Healy. Despite both Healy's and the wolverines' correspondingly "mindless" expressions of violence, the real villain of Ellroy's work is a ruthless and detached post-war industrial culture that transforms both animal and human life into useable commodities. Animals in particular are frequently situated as victims of monstrous human behaviours over the course of the Quartet, raising broader questions about the stability of the human/animal binary. In the same way that Ellroy's work ultimately operates to destabilise the logic that underpins the oppressive practices of racist white cops, the hyperbolic human violence that unfolds similarly destabilises and deconstructs one of the key aspects of hierarchical species difference; namely the separation between the civilised and the savage.

"Werewolf Murder": Identity, Monstrosity and "Otherness" in *The Black Dahlia*

The Black Dahlia is the opening text of the *L.A. Quartet* and prefigures the explicit intersections between animalism and varying taxonomies of deviancy that permeate Ellroy's succeeding volumes. The text centres around the real-life, unsolved 1947 murder of aspiring actress Elizabeth Short, whose eviscerated remains were discovered stripped and abandoned on a vacant lot on South Norton Avenue, Los Angeles. The case holds a long-standing fascination for Ellroy, one that is intimately bound up with the unsolved murder of his own mother Geneva Hilliker. In his memoir *My Dark Places*, Ellroy describes how the "Black Dahlia" case operated as a symbolic substitute for his mother's death, one that allowed him to construct "strong, narrative based fantasies" in which he rescued Elizabeth Short. Ellroy describes how these vivid imaginings would invariably involve him rescuing Betty "and becom[ing] her lover", or "tracking down her killer and executing" him in passionate revenge.[16]

At the thematic epicentre of Ellroy's novel is the gruesome iconography of Elizabeth Short's brutalised body, chiefly its spectacular commodification via the mass media. Following the discovery of her remains, Short becomes the focus of intense media scrutiny and circulation, the gruesome details of her evisceration regurgitated and fetishised for commercial, and in the case of District Attorney Ellis Loew, political gain. The inexplicable and brutal nature of the murder makes it a breeding ground for crazed fantasies of monstrosity, foregrounding the *L.A. Quartet*'s deeper anxieties regarding violent crime and forms of "otherness". This becomes explicitly articulated via the media's focus on the fantasy "werewolf" killer, a symbol of demonic violence that becomes implicitly bound up with the text's codifications of race and degeneracy.[17] The symbol of the "werewolf psychopath" becomes expressive of the text's deeper racial divisions, particularly the continued conflation of blackness with criminality and other monstrous behaviours. This is made apparent in the early stages of the text, when protagonist Detective "Bucky" Bleichert's search for rampaging rapist Junior Nash leads him into contact with degenerate and violent "nigger youth gangs" (57). Known associates of Junior Nash, these gangs come to the attention of the LAPD due their predilection for violence and animal cruelty, particularly the torturing and then disposal of "chopped-up dead cats into the cemeteries of Santa Monica and Gower" (57). The conjunction is unambiguous here, as articulations of animalism become literally manifested in brutal acts of violence perpetrated against animals, a pattern that continues throughout the *L.A. Quartet*. Such representations are frequently energised by a sense of paranoia born out of a perceived threat posed by demonic and deviant races and sexualities.

Crucially, Ellroy's Los Angeles is a space brutally schematised by categories of identity, particularly race. This is forcefully emphasised in an extract from *The Black Dahlia*. Whilst out on patrol, Bucky Bleichert's description of the physical and ethnic topography of the "Newton Street Division" provides a clear indication of how race operates in the *L.A. Quartet*:

> Newton Street Division was southeast of downtown LA, 95 percent slums, 95 percent negroes, all trouble. There were bottle gangs and crap games on every corner; liquor stores, hair straightening parlours and poolrooms on every block, code three calls to the station twenty-four hours a day. The local winos drank "Green Lizard"—cologne cut with Old Monterey white port, and the standard pop for a whore was one dollar, a buck and a quarter if you used "her place"—the abandoned cars in the auto graveyard at 56th central.[18]

Depicting an impoverished and dilapidated social space permeated with instances of gambling, prostitution and degeneracy, it is race that becomes the constituting factor in such stark portrayals of social disunity. A mode of "categorisation that affirms and reinforces existing social and economic inequalities", blackness emerges as what Andrew Pepper describes as a "signifier of mindless savagery and ultimate worthlessness which operates within 'official' linguistic structures to secure and legitimise white hegemony".[19] Pepper points to the various mechanisms of institutional racism in the *L.A. Quartet* as a central facet in the perpetuation of these paradigms of power. It is no surprise then that Bucky's unremittingly violent policing of the various criminal factions of Newton Street earns him respect and notoriety within the division. Bucky's brutal patrolling of impoverished black communities is justified precisely because of an institutionalised logic that positions criminality and blackness as indivisible. Such a logic is emphasised when Bucky ruthlessly breaks up a "rigged card game" with his baton, an action that is justified by his assertion that "some people don't respond to civility".[20] The implication here is explicit and reasserts a perception of blackness as an "uncivilised animalistic force", one that represents a threat to the hegemony of white male power.[21]

Nonetheless, the conclusion to the *Black Dahlia* operates to destabilise these assumptions about criminality and deviancy and, in the process, highlights the complexities and paradoxes of identity that underlie Ellroy's work. Far from being the work of a perverse and deranged werewolf killer, Elizabeth Short's death is ultimately connected to the affluent and politically reputable Sprague family, whose outward projection of respectability masks a warped history of incest, depravity and exploitation. From the rotting foundations upholding the properties from which Emmet Sprague made his millions, to the brutal depletion of Elizabeth Short's body at the hands Ramona Sprague and her deranged lover George Tilden, this poisonous family are a warped and hyperbolised projection of the violence and derangement percolating behind various white institutions of power. As Pepper suggests, the "luridly bigoted tendencies" of Ellroy's white cops combined with the degeneracy and psychosexual deviancy of his power-hungry politicians and dysfunctional white entrepreneurs, ultimately function to erode the logic that whiteness is any more civilised than the (often racially other) criminal classes they seek to control.[22]

By presenting the devastating corruption at the centre of these mechanisms of power, this moment of revelation disrupts and threatens the logic that energises the orientalist discourse that has historically positioned whiteness as superior to that of the uncivilised, deviant racial other. Whilst, on the one hand, Ellroy seems to propagate a brand of misogynistic and violent, institutionalised whiteness—one that has come to typify the hard-boiled novel historically—his unremitting portrayal of the savagery and violence of white characters simultaneously destabilises entrenched paradigms of otherness that exclusively situate black and homosexual characters as violent and deviant. As Tim Ryan suggests, Ellroy's work not only operates to make "whiteness visible" and "show how power operates", but also to "deconstruct the discourses that naturalize that power". Thus, through a lurid confrontation with the criminal and appalling acts undertaken by white males, the reader is confronted with, and made complicit in, the "savage exploitation, oppression and violence that such adventures entail".[23]

Although ultimately a myth, the "werewolf killer" actualises these patterns of moral panic regarding the threat of an uncivilised, devouring other, whilst also prefiguring the persistence of animal–human hybrids that will continue to punctuate Ellroy's work; the most disturbing example of which are the mutilated victims of serial killer Loren Atherton in *L.A. Confidential.* Dubbed a "monster" and "Dr Frankenstein" by newspapers following a string of child abductions and murders, paedophile Loren Atherton's modus operandi involves "killing and building hybrid children".[24] Not only does Atherton splice limbs and other body parts from victim to victim, he also mutilates birds and then attaches the wings to complete his "creations". Combined, both the fantasy "werewolf killer" and Atherton's monstrous mutilations can most obviously—as mentioned—be seen to reinscribe the "familiar suppositions" about the "desiring and unregulated body" that underscore many "essentialist views of categories of difference".[25] In other words, both superficially reinscribe the "animal" as a generic signifier for violent cravings and transgressive acts. Yet, it is also possible to see these hybrid articulations as indicative of a broader crisis of identity permeating the *L.A. Quartet*, not simply regarding questions of gender, sexuality and race, but also in relation to the human subject more broadly. The hyperbolic "inhuman" violence perpetrated by humans in Ellroy's texts—often even against animals themselves—operates to destabilise human–animal binaries, whilst simultaneously threatening our very "definitions of the human itself".[26]

The "Wolverine Monster": Destabilising the Human–Animal Binary

The tensions and contradictions surrounding classifications of identity, criminality and monstrosity reach their apotheosis in the second volume of the *L.A. Quartet*, *The Big Nowhere*, as do these broader tensions regarding the animal/human. The text converges around two principle plot lines. The first of these follows the implementation of a new federal task force, one that seeks to investigate rumours of communist activity and governmental subversion within the Hollywood unions. This runs concurrent with the L.A. Sheriff Department's investigation into a string of violent, homosexual sex murders, after the mutilated body of local Jazz musician Marty Goines is discovered dumped on a weed-strewn vacant lot. The first victim of the predatory killer dubbed the "Wolverine Monster"—due to the perpetrator's predilection for biting victims whilst wearing dentures with wolverine teeth attached—the brutal, animalistic violence inflicted upon Goines's body prefigures a pattern of conjunction between homosexuality and monstrosity that permeates *The Big Nowhere*. On arriving at the crime scene, Detective Danny Upshaw is confronted by a mutilated body covered with:

> six ovals measured to within three centimetres of each other. They all bore teethmark outlines too shredded to cut a cast from—and all were too large to have been made by a human mouth biting straight down. The killer used an animal or animals in the post-mortem abuse of his victim.[27]

This depiction of extreme violence inflicted upon gay men operates as a hyperbolised portrayal of "queer" sub cultures more generally, a logic that perceives "homosexuals as foreign, subhuman, monstrous and animalistic".[28] Such violence is rendered as the logical culmination of a form of deviant and dangerous sexuality and subsequently functions as an extreme manifestation of a broader pattern of exchange between atavistic behaviours and homosexual identity.

The murder of Goines is the first in a number of "queer slashes", extreme and violent murders that are nonetheless a low priority for the LAPD due to the perceived deviancy of the victims.[29] The wolverine killer thus emerges as a hyperbolised manifestation of a form of sexual "otherness", one that is rigidly codified as predatory, dangerous and violent and thus a threat to established hierarchies of control. The wolverine killer is eventually revealed as Coleman Healy, the disaffected and illegitimate son of blacklisted actor Reynold Loftis, who in turn is one of the key figures under scrutiny in the grand jury investigation into communist activities in Hollywood. After witnessing the murder of Jose Diaz at the hands of arch villain Dudley Smith, Healy's past becomes intimately entrenched in larger criminal manoeuvrings linking the LAPD to organised crime. The subsequent cover up of the infamous "Sleepy Lagoon Murder" drives Healy into the arms of the father who previously rejected him.[30] Reynolds employs plastic surgeon Terry Lux to alter Coleman's appearance so that it resembles his own, allowing his son to elide the violent repercussions of Dudley Smith. The surgery precipitates a perverse, incestuous affair between father and son. The narcissistic Loftis becomes "so enthralled by his on-screen image", that he uses his estranged son to enact a "fantasy love affair with himself".[31] After Healy is eventually rejected by his father, Lux alters his face again, transforming him unrecognisably from his previous look. Disaffected and seeking revenge, Healy disguises himself as

Loftis and carries out a string of sex killings to implicate his father in the murders. In doing so he "emulates the violence of the wolverines—creatures whose ferocious savagery he has come to admire".[32]

Healy's obsession with the wolverine has a long and disturbing genealogy, one that can be traced back to his love of an old Victrola record called "Wolverine Blues". Disaffected after being rejected by the army, Healy develops a violent and sexual obsession with animals, one that culminates in his discovery of "strange-o" Thomas Cormier's wolverine enclosure. After a period of secretly visiting and feeding the animals, one-night Healy is seized by "an incredible urge to pet and hold one", only to fall victim to a violent attack.[33] In the ensuing struggle, Healy kills and then skins the wolverine, consuming its raw flesh, making dentures out of its teeth. In an effort to understand the brutal savagery of Healy's murders, detective Danny Upshaw visits Cormier's wolverine enclosure, an encounter that rigidly frames Healy's violence as an expression of something "animalistic" and "other":

> Danny found a good sun angle—light square on a middle pen; he squatted down and looked, his nose to the wire. Inside, a long creature paced, turning in circles, snapping at the walls. Its teeth glinted; its claws scraped the floor; it looked like a coiled muscle that would not stop coiling until it killed and slept in satiation—or died. Danny watched, feeling the beast's power, feeling HIM feeling it. Cormier talked. "*Gulo luscus* is two things; smart and intractable. I've known them to develop a taste for deer, hide in trees and toss nice edible bark down to lure them over, then jump down and rip the deer's jugular clean to the windpipe. Once they get a whiff of blood, they will not stop persisting. I've heard wolverines stalking cougars wounded in mating battles. They'll jab them from behind, take nips out and run away, a little meat here and there until the cougar nearly bleeds to death. When the poor fellow's dead, *Gulo* attacks frontally, claws the cougar's eyes out of his head and eats them like gumballs".[34]

Ellroy's depiction of the wolverines further expounds the aforementioned political undercurrents of animality as a broad signifier of various forms of transgressive, threatening "otherness". With their almost cartoonish glinting teeth and scraping claws, this generalised animality typifies what John Miller describes as an "anthropocentric" rendering of "the animal", one that "strategically exaggerates the role of violence in animal behaviour".[35] Such savage animality, as incarnated by the wolverines, becomes a means of framing Coleman's brutal and sexualised violence as something decidedly atavistic and "non-human". For Paul Shepard, the exaggerated, bloodthirsty viciousness of Ellroy's wolverines is indicative of the way in which animals repeatedly operate as "handles for abstractions" in the traditions of storytelling, a way to comfortably re-orientate and embody that which is unknown or inexplicable.[36] In this case, the abstraction of Healy's hyperbolic violence is re-routed into the figure of the wolverine, ultimately validating the Quartet's continued convergence of animality with various forms of monstrous "otherness".

Yet, whilst *The Big Nowhere* seemingly frames such murderous aggression as intrinsic to animal nature, both Healy and the Wolverines are similarly victims of a violently stratified and brutally alienating post-boosterism Los Angeles, one driven by an individualistic, capitalist agenda that indiscriminately commodifies and oppresses. As with the exploitation and confinement of the wolverines at the hands of Cormier, Healy's so-called animal savagery is the consequence of a comparable form of pervasive and institutionally entrenched human brutality that saturates not only *The Big Nowhere*, but the *L.A. Quartet* more broadly. From police corruption to the warped aspirations of Hollywood film, Jim Mancall argues that Healy is "ultimately powerless before the violence he has suffered at the hands of others; someone made him".[37] Thus, rather than reading the wolverines as emblematic of an intrinsic animality, it becomes possible to see their subjugation and abuse— at the hands of Healy and Cormier—as a direct consequence of a more extensive and pernicious system of human violence, a system that ultimately complicates the rigid human/animal binary the text seems to purport. As much "victim as victimizer", Healy emerges—according to Mancall—as another one of Ellroy's hybrids, a "fragmented personality, part human and part animal".[38] Like his codifications of homosexuality and race, Ellroy's framing of animality is fuelled by challenging contradictions. Whilst the *L.A. Quartet* seems to proliferate a generic concept of animality that aligns transgressive identities and behaviours with the atavistic and non-human, the extent of the human violence that saturates Ellroy's Los Angeles operates to destabilise the inviolability of the human subject, particularly the opposition between civilised/uncivilised. In *The Big Nowhere* in particular, the "natural" violence of the wolverines is projected against the backdrop of unfathomable human corruption and cruelty that manifests itself on an institutional level.

These larger questions that the text raises concerning identity and the definition of the human subject are played out on a microcosmic scale via Detective Danny Upshaw's homosexual crisis. In *The Big Nowhere*, institutional anxieties regarding the threat of homosexuality are continually "played out" via what Josh Meyer describes as "a series of symbolic associations", particularly between homosexuality, communism, incest, animalism and crime.[39] For Meyer, this becomes specifically expressed via the theme of "contagion", the discourses of which punctuate the text on multiple levels. As Upshaw's investigation into the sex killings draws him deeper into the homosexual subculture of the city, he is progressively forced to confront his own repressed sexuality. His immersion in, identification with, and attraction to this "deviant" sexual underworld is intimately bound up with a broader narrative logic of contamination. Upshaw's inability to suppress his secret becomes most explicitly and forcefully expressed through his use of the "Man Camera". Described as an investigative tool developed by criminologist Hans Maslick, the technique involves recreating crime scenes cinematically by "screening details from the perpetrator's point of view" using actual camera angles and shots. Through this psychological projection, the investigator's eyes become a "lens capable of zooming in and zooming out, freezing close ups, and selecting background motifs to interpret crime scene evidence in an aesthetic light".[40] The "Man Camera" is theoretically designed to function as an objective, documentation device, an extension of Upshaw's obsession with forensic methodology and scientific observation. Yet, as Upshaw is drawn deeper into the homosexual subculture of the city, he begins to experience "Man Camera malfunctions", losing control of the device and consequently his vision:

> Danny pressed his face to the window and looked in.
>
> That close he got distortion blur, Man Camera malfunctions. He pulled back so that his eyes could capture a larger frame, saw tuxedos entwined in movement, cheek-to-cheek tangos, all male. The faces were up against each other so that they couldn't be distinguished individually; Danny zoomed out, in, out, in, until he was pressed into the window glass with pins and needles localised between his legs, his eyes honing for mid-shots, close-ups, faces.
>
> More blur, blips of arms, legs, a cart being pushed and a man in white carrying a punch bowl. Out, in, out, better focus, no faces, then Tim and Coleman the alto together, swaying to hard jazz.[41]

Upshaw's ontological crisis becomes increasingly channelled through these pulsing Man Camera blips, a disorienting interplay of fantasy, reality and memory merged with the pornographic violence of crime scenes. Ungoverned, the Man Camera invokes flashes of eviscerated innards, dismembered penises and other mutilations in "wraparound technicolour", projecting Upshaw's warped and tormented identification with sexual violence. Here Upshaw's loss of visual agency becomes symptomatic of his impotent attempts to exert "a masculine vision".[42] This veritable crisis of the male gaze signifies a crisis of masculine authority more generally and emerges as a direct consequence of Upshaw's inability to reconcile his own latent homosexual desires. The embedded logic of contagion, one that Meyer argues is central to *The Big Nowhere*'s narrative patterns, is certainly palpable in Upshaw's contact with, and ultimate identification with, the text's homosexual subculture. Upshaw becomes progressively "infected" by his proximity and exposure to this underworld, suffering disorientation, panic attacks and nightmare hallucinations. His experience voyeurising the gay orgy conjures forth a number of warped visions that reinforce the texts inscription of homosexuality with attendant ideas of violence and monstrosity. Disturbing, kaleidoscopic projections obstruct Upshaw's objective perception, transforming the spaces of the city into a gory, violent nightmare and passing civilians into deformed "gargoyles".[43] Ultimately, Upshaw's infection by the "queer" subculture of the city and subsequent loss of masculine visual agency leads to his demise. Fearful that his fantasies and desires will be exposed to the public, Upshaw eventually commits suicide, slitting his throat with a serrated-edged carving knife.

Upshaw's fractured subjectivity ultimately becomes reflective of the broader crisis of the "human subject" in the *L.A. Quartet.* For all the savagery and brutality that becomes associated with literal and figurative animals, once again it is ultimately the monolithic mechanisms of power—ones energised by an ideology of white superiority—that become most implicated in the text's criminal conspiracies. From the crazed containment plans of the murderous Dudley Smith to the deceitful political strategies of district attorney Ellis Loew, Ellroy's work problematises assumptions concerning animality and "otherness". As Jim Mancall suggests, *The Big Nowhere* exposes the fact that "grand ideologies such as democracy and liberty are mobilized by those in power so as to protect their own advantages and that these terms often hide violence, exploitation and conspiracy".[44] As a consequence, the hypocrisy of the racism, homophobia and general discriminatory practices of the police become brutally visible.

Ellroy's texts project a brutal vision of post-war Los Angeles, one in which animality superficially becomes tied not only to criminality, but to various marginalised cultures, groups and individuals. Yet, Ellroy's work continues to be wrought by challenging contradictions. In the same way that Ellroy's exposure of the violence and corruption at the centre of white characters and white institutions subverts the logic that conflates "otherness" with violence and deviancy, the level of human perversion evident in the *Quartet* undermines and weakens the logic that underlies the human–animal binary. That is not to say that Ellroy goes all the way towards dislodging these entrenched hierarchies. Despite the nuances explored in this essay, the continued persecution and marginalisation of the "other", combined with the lack of retribution for those associated with the mechanisms of white social power, makes it difficult to suggest that Ellroy entirely shatters such inequalities. In fact, such an unyielding portrayal of "bad white men" and could in fact been seen to glorify and thus further cement such entrenched binaries of power, including the human–animal duality.[45] In either case, Ellroy's work offers a multifaceted and complex interrogation of racial, sexual and animal politics and will no doubt continue to polarise academic and popular opinion.

Notes

1. Mancall, *Companion*, 1.
2. This use of animals extends beyond Ellroy's in fiction. In his 2011 television show "City of Demons", Ellroy was accompanied on screen by a CGI animated bull terrier called Barko, the "corrupt police dog".
3. Ellroy, "The Choirboys", 350.
4. Powell, "Coda", 173.
5. The *L.A. Quartet* is comprised of *The Black Dahlia* (1987), *The Big Nowhere* (1988), *L.A. Confidential* (1990) and *White Jazz* (1992).
6. Scaggs, *Crime Fiction*, 99.
7. Pittard, *Purity and Contamination*, 110.
8. Thomas, *Detective Fiction*, 220.
9. Morrison, *Playing in the Dark*, 5 quoted in Kennedy, "Black Noir", 224.
10. Chandler, *Big Sleep*, 109.
11. Meyer, "Scarlett Fever", 41.
12. Derrida, *The Animal*, 162.
13. Waldau, *Animal Studies*, 2.
14. Waldau, *Animal Studies*, 2.
15. Armstrong, *What Animals Mean*, 3.
16. Ellroy, *Dark Places*, 111.
17. Ellroy, *Black Dahlia*, 92.

18. Ellroy, *Black Dahlia*, 299.
19. Pepper, *Contemporary American*, 43.
20. Ellroy, *Black Dahlia*, 302.
21. Ghaill, *Understanding Masculinities*, 134.
22. Pepper, *Contemporary American*, 45.
23. Ryan, "Shiny Bleach Job", 273.
24. Ellroy, *L.A. Confidential*, 266.
25. McKay and Miller, *Werewolves*, 5.
26. Armstrong, *What Animals Mean*, 67.
27. Ellroy, *Big Nowhere*, 9.
28. Plummer, *One of the Boys*, 33.
29. Ellroy, *Big Nowhere*, 370.
30. The death of José Gallardo Díaz—the so called Sleepy Lagoon Murder—is one of many instances of "true crime" that Ellroy seamlessly integrates into his fictional reimagining of post-war Los Angeles. Ultimately blamed on a group of Mexican-American youths, these convictions are often perceived as the catalyst for the ensuing "Zoot Suit Riots" of 1943 (as fictionalised in *The Black Dahlia*). Following much protestation about the LAPD's handling of the case—particularly the lack of evidence used to indict the youths—the convictions were eventually repealed in 1944.
31. Sunderland, "Revisiting Paranoia".
32. Mancall, *Companion*, 104.
33. Ellroy, *Big Nowhere*, 456–61.
34. Ellroy, *Big Nowhere*, 354–355.
35. Miller, *Empire*, 167.
36. Shepard, *Thinking Animals*, 117.
37. Mancall, *Companion*, 32.
38. Mancall, *Companion*, 31.
39. Meyer, "Scarlett Fever", 40.
40. Ellroy, *Big Nowhere*, 94.
41. Ellroy, *Big Nowhere*, 168.
42. Cohen, *Spectacular Allegories*, 133.
43. Ellroy, *Big Nowhere*, 95.
44. Mancall, *Companion*, 34.
45. Abbott, *The Street was Mine*, 194.

Works Cited

Abbott, Megan. *The Street Was Mine: White Masculinity in Hardboiled Fiction and Film Noir.* New York: Palgrave Macmillan, 2002.

Armstrong, Philip. *What Animals Mean in the Fiction of Modernity.* Abingdon: Routledge, 2008.

Chandler, Raymond. *The Big Sleep*. London: Penguin, 2011.
Cohen, Josh. *Spectacular Allegories: Postmodern American Writing and the Politics of Seeing* . London: Pluto Press, 1998.
Derrida, Jacques. *The Animal That Therefore I Am*. Edited by Marie-Louise Mallet. Translated by David Wills. New York: Fordham University Press, 2008.
Ellroy, James. "'The Choirboys." In *The Best American Crime Writing, 2005*, edited by Otto Penzler, James Ellroy, & Thomas H Cook. New York: Harper Perennial, 2005.
Ellroy, James. *L.A. Confidential*. London: Windmill, 2011a.
Ellroy, James. *My Dark Places*. London: Windmill Books, 2010.
Ellroy, James. *The Big Nowhere*. London: Windmill, 2011b.
Ellroy, James. *The Black Dahlia*. London: Windmill, 2011c.
Ghaill, Máirtín Mac an. *Understanding Masculinities: Social Relations and Cultural Arenas*. London: Open University Press, 1996.
Kennedy, Liam. "Black Noir: Race and Urban Space in Walter Moselys Detective Fiction." In *Diversity and Detective Fiction*, edited by Kathleen Gregory Klein, 224–244. Bowling Green: Bowling Green State University Popular Press, 1999.
Mancall, Jim. *James Ellroy: A Companion to Mystery Fiction*. Jefferson: McFarland, 2014.
McKay, Robert, and John Miller. *Werewolves, Wolves and the Gothic*. Cardiff: University of Wales Press, 2017.
Meyer, Joshua. "Scarlet Fever: Communism, Crime and Contagion in James Ellroy's The Big Nowhere." *Clues* 33, no. 1 (2015): 40–50.
Miller, John. *Empire and the Animal Body: Violence, Identity and Ecology in Victorian Adventure Fiction*. London: Anthem Press, 2014.
Morrison, Toni. *Playing in the Dark: Whiteness and the Literary Imagination*. Cambridge: Harvard University Press, 1992.
Pepper, Andrew. *The Contemporary American Crime Novel: Race, Ethnicity, Gender, Class*. Edinburgh: Edinburgh University Press, 2000.
Pittard, Chris. *Purity and Contamination in Late Victorian Detective Fiction*. Farnham: Ashgate, 2011.
Plummer, David. *One of the Boys: Masculinity, Homophobia, and Modern Manhood*. Abingdon: Routledge, 2016.
Powell, Steven. "Coda for Crime Fiction." In *Conversations with James Ellroy*, edited by Steven Powell, 169–175. Jackson: University Press of Mississippi, 2012.
Ryan, Tim. "'One Shiny Bleach Job': The Power of Whiteness in James Ellroy's American Tabloid." *The Journal of American Culture* 27, no. 3 (2004): 271–278.
Scaggs, John. *Crime Fiction*. Abingdon: Routledge, 2005.
Shepard, Paul. *Thinking Animals: Animals and the Development of Human Intelligence*. London: The University of Georgia Press, 1978.

Sunderland, Maureen. "Revisiting Paranoia: The Witch-Hunts in James Ellroy's The Big Nowhere and Walter Mosley's A Red Death." In *Millenial Detective: Essays on Trends in Crime Fiction, Film and Television*, edited by Malcah Effron, 142–156. Jefferson: McFarland, 2011.
Thomas, Ronald R. *Detective Fiction and the Rise of Forensic Science*. Cambridge: Cambridge University Press, 1999.
Waldau, Paul. *Animal Studies: An Introduction*. Oxford: Oxford University Press, 2013.

ETHICS

The Psittacine Witness: Parrot Talk and Animal Ethics in Earl Derr Biggers' *The Chinese Parrot* and Earl Stanley Gardner's *The Case of the Perjured Parrot*

John Miller

An 1845 court case in Dublin featured a remarkable witness. When a Mr Moore was accused of stealing a parrot from a Mr Davis, it seemed logical to ask the bird in question for his side of the story. *The Mirror* provided a detailed account of the parrot's testimony, gleefully noting that the prosecution's star witness raised certain logistical challenges. A solicitor acting for the defence asked of the bird, "Do you intend to have him sworn? If you do, I'd like to know which book, as it is very likely he is a heathen or a Turk".[1] The case pivoted on Mr Davis's success in being able to prompt responses to a series of questions about animal noises ("Tell me, old fellow, what does the dog say?") Such was the parrot's performance that *The Mirror* even noted "there appeared to be a touch of waggery in the action

J. Miller (✉)
University of Sheffield, Sheffield, UK
e-mail: John.Miller@shef.ac.uk

R. Hawthorn, J. Miller (eds.), *Animals in Detective Fiction*, Palgrave Studies in Animals and Literature,
https://doi.org/10.1007/978-3-031-09241-1_5

of the animal".[2] The journalist's amusement at the bird's antics indicates the comic effects that parrots are likely to produce in the august setting of the courtroom. Yet, for all the humour in *The Mirror*'s report, the idea of interrogating the parrot witness takes us to the heart of conventional conceptions of human–animal difference. Did the parrot know what he was saying? Might he even be capable of something approaching rational thought? As we shall see, parrot talk has long exercised philosophers concerned with the examination of the exceptional status of the human. The psittacine witness is a being burdened with some weighty philosophical issues.

This chapter examines the representation of parrot witnesses in two American detective novels which engage not just with the theme of animal ontology—the parrot's apparent status on the cusp of rationality and brutality—but with the ethical questions that are attached to the imagining of parrots before the law. In both Earl Derr Biggers' Charlie Chan story *The Chinese Parrot* (1926) and Erle Stanley Gardner's Perry Mason novel *The Case of the Perjured Parrot* (1939), the resolution of the mystery depends on the detective's ability to interpret the utterances of a parrot. In Biggers' novel Chan is brought from his home in Hawai'i to the American mainland to help with the sale of a string of pearls by the impecunious socialite Sally Jordan, via the jeweller Alexander Eden, to the mega-rich financier P. J. Madden who intends to give them to his daughter Evelyn. As Chan arrives incognito at Madden's ranch in the California desert along with the jeweller's son Bob Eden to oversee the transfer of the pearls to their new owner, a "horrible cry" shatters the night: "'Help! Put down that gun! Help! Help!'"[3] The sound is no human voice, but the cry of the novel's eponymous parrot, a bird named Tony who announces the apparent crime that the subsequent plot unravels. Comparably, in Gardner's *The Case of the Perjured Parrot*, a parrot is discovered in a mountain cabin next to the murdered body of the "eccentric multimillionaire" Fremont C. Sabin.[4] For Mason the case comes to seem one "which entirely revolves around a parrot" (92). Here is the bird's riddle: "Put that gun down, Helen! Don't shoot! Squawk. Squawk. My God, you've shot me!" (47). If this parrot is somewhat more explicit than Tony in the evidence he communicates, with two Helens in Gardner's cast of characters it is far from an open-and-shut case. Moreover, as the mystery deepens, it is clear that there is more than a single parrot with its beak in the case.

Evidently, the narrative patterning of these novels casts the parrot in a particular formal role. A genre as structurally overdetermined as classic detective fiction invites the reading of all subjects as plot devices: the witness, the victim, the detective, etc. There is some ambivalence, however, about how the role of the parrot might be most accurately described. As Bruce Boehrer argues, technically parrots are "evidence not witnesses" since they are usually thought to be unable to take the oath required of them to take the stand.[5] Yet, they are also more than evidence; at least, as Boehrer continues, "a living, talking creature tends to look a lot more like a witness than like ordinary evidence".[6] Accordingly, while in both cases it is primarily the words the birds utter that drive the plot and test the ingenuity of the detective, each of these texts also pays attention to the bodies of the parrots, particularly in the depiction of the untimely ends for which the birds are destined. *The Chinese Parrot* and *The Case of the Perjured Parrot* are not, as I explain below, explicitly pro-animal texts, but nonetheless their attention to their parrot characters facilitates a partial critique of a conventional anthropocentric politics.

Since Biggers's and Gardner's parrots derive part of their meaning from the role of talking birds in a wider cultural and intellectual history, I offer a brief initial survey of the role of parrots in Enlightenment philosophy (a key context for the detective form, as we have seen in the introduction to this volume) before examining the specific ethical implications of *The Chinese Parrot* and *The Case of the Perjured Parrot*. In *The Chinese Parrot*, animal ethics operate in close relation to the novel's progressive, if at times compromised, racial politics in a way that encourages sympathy, albeit somewhat unreliably, for the characters, both animal and human, consigned to the margins of the novel. *The Case of the Perjured Parrot*, meanwhile, in developing its intrigue through a number of different parrots raises questions about fungibility (the idea that one parrot can be straightforwardly exchanged for another parrot) in order to foreground the problematic logic of the animals as commodity. Ultimately, global capital represents the overarching context in which ethical relationship with parrots unfold. In conclusion, I argue that in both texts this economic aspect provides a notable tension with and counterpoint to the ethical commitments the parrots invite.

Enlightenment Parrots

Debates around the philosophical status of talking birds are recurrent and weighty enough to render parrots as something like a cornerstone of Enlightenment humanism. Most notably, parrots make a brief, but telling appearance in René Descartes' *Discourse on Method* (1637). In wrestling with the significance of parrot talk, Descartes concludes that "parrots can utter words like ourselves, and are yet unable to speak as we do, that is, so as to show that they understand what they say".[7] From this interpretation of the meaning of parrot vocalisation, Descartes reaches the further conclusion: "not only that the brutes have less Reason than man, but that they have none at all".[8] John Locke, by contrast, found cause to look more generously on the vocal facility of parrots. An account of a talking parrot from Brazil enables Locke to "countenance the supposition of a rational *parrot*" (original emphasis), though this concession does not necessitate redrawing the barrier between man and brute.[9] The point Locke makes—in contradistinction to the Cartesian view—is that "it is not the idea of a thinking or a rational being alone that makes the *idea* of a *man* in most people's sense, but of a body so and so shaped".[10] However smart a parrot might be, its physical form expresses its inferiority to humans.

If the parrot in Descartes' and Locke's accounts ultimately reaffirms a coherent sense of human pre-eminence, the question of a parrot's location in the chain of being is not always so clear-cut. As Thomas DiPiero summarises, in Enlightenment thought talking birds came to symbolize "the irresolution" of the debate about whether the world was "strictly material" or divided, as Descartes thought, "into thinking and nonthinking matter".[11] Both sides "could press them into service to show either that humanity occupies a privileged position in the order of things [...] or that people and animals are fundamentally the same".[12] In the eighteenth century, for instance, Julien Offroy de La Mettrie returned to Locke's argument about the Brazilian parrot to make a case for the potential of non-human creatures. Moving from parrots to apes, La Mettrie states that "I hardly doubt at all if this animal were perfectly trained, we would succeed in teaching him to utter sounds and consequently to learn a language. Then he would no longer be a wild man, nor an imperfect man, but a perfect man".[13] The effect of parrot talk in La Mettrie's controversial account is to generate a position in which human exceptionalism is increasingly unstable.

Even if talking birds, in many cases, ultimately come to reinforce an anthropocentric position, they also challenge ontologies of human–animal difference in a way which has a bearing on how we think about ethical obligations towards animals. A culture that sees difference hierarchically inevitably produces ethical priorities in relation to the terms of that difference. The use of language (to return again to a Cartesian logic) indicates the possession of reason which conventionally signals a being deserving of ethical consideration. Consequently, the ability to talk retains an ethical dimension within an anthropocentric value system that sees resemblance to the human as part of the basis for ethical thinking. That said, there have been significant movements away from a model that bases ethical consideration on the manifestation of cognitive functions characterised as human or near-human. Jeremy Bentham famously redirected the focus of animal philosophy—in a phrase of unusually literal relevance to parrots— by suggesting the crucial question is not "'Can they reason?' or 'Can they talk?' but 'Can they suffer?'[14] Bentham's influential dictum has proved a persistent touchstone in pro-animal movements as something like a manifesto point; moreover, it is the question of suffering around which a vast proportion of scholarly interventions in the cultures and politics of human–animal relationships revolve. As Bentham sidelines the importance of cognitive abilities in favour of physical sensitivity, he reduces the significance of parrot talk for wider debates about ethics. Even so, the idea of animal language is not without ethical force. Voice—as the faculty of speech—is widely used metonymically as a figure for political agency. And there is good reason why the possibility that parrot talk might meaningfully express some kind of interiority has proved hard to erase. Research in animal cognition—particularly in relation to an African Grey called Alex—has shown that rather than just mimicking human sounds, parrots can appear capable of using "language creatively".[15] The question of ethical obligations towards parrots gains a particular inflection therefore through their facility of "human" speech.

Accordingly, the representation of the parrots in Biggers's and Gardner's novels exists within a conventional humanist ethical framework that sees no problem in general with violence against animals. *The Chinese Parrot* features animal death as an unquestioned part of the *mise-en-scène* of the Californian desert. Bob Eden and the novel's love interest Paula Wendell are brought together initially in a moment of meat comedy in the local inn when the tough steak he orders skids off his plate and drops "on to the knees of the girl" (42). Later in a casual aside, the town's newspaper editor Holley, one of Chan's key allies, notes that "coyotes are getting pretty bad in the valley" and a number of ranchers are "putting out traps" (74). *The Case of the Perjured Parrot* offers a little more in term s of feelings for the wider animal world, albeit in a sentimental manner. Mason and his secretary Della Street find themselves drawn to the wildlife around Sabin's cabin: Street makes friends with the chipmunks while Mason feeds the blue jays (30–31), though their attention is quickly redirected to the central human action and, as with Biggers' novel, there are plenty of off-hand references to meat that affirm the novel's anthropocentric structures. In one jokey aside, parrots are included in the frame of edibility in a way that emphasises the general assumption about the status of animals. Mason's associate Paul Drake pauses to ponder the best way to eat parrot: "Make a parrot pie, or do you broil 'em on toast?" (16) If parrots produce ethical responses, this element of the texts emerges through a complex network of commitments and interests rather than as the expression of a straightforward pro-animal agenda.

Grief and Race in *The Chinese Parrot*

Despite the novel's title, Tony is not Chinese at all. Rather, he is a "little gray Australian bird that some sea captain gave Madden several years ago". Before taking residence at the millionaire's ranch, Tony "used to hang out in a barroom on an Australian boat" (54). Madden acquired Tony, according to Holley, as "company for the old caretaker", Louie Wong; consequently, we learn, Tony "has learned to speak Chinese", rattling "along in two languages. A regular linguist" (54). Tony acquires his sobriquet "the Chinese parrot", Holley informs us, from the neighbouring ranchers (54). The key element in Biggers' introduction of this cosmopolitan bird is evidently the connection between Wong and Tony: The parrot picks up Chinese "from associating with Louie" (54), a link which develops wider significance in the context of the representation of the novel's several Chinese-American characters.

The association between Tony and Wong emerges most dramatically in the depiction of their respective demises. To pick up the story of *The Chinese Parrot*: as Bob Eden and Charlie Chan take up residence in Madden's ranch just outside the desert town of Eldorado, the detective is now disguised as the Chinese cook Ah Kim with the pearls concealed around his waist. Eden is charged with the task of handing over the pearls to Madden until the parrot's cry gives them pause and they begin a series of delaying tactics. With no corpse or missing person reported, the initial mystery is not just a question of discovering the murderer, but also of discovering the victim. As it turns out, no one has been murdered (not to start with anyway). The first murder victim is none other than Tony, killed presumably because he knew too much (or at least had too much to say). The finger of suspicion points squarely at Madden's secretary Thorn who is unable to disguise the "ghost of a smile" when Chan (as Ah Kim) reveals the parrot's death. Together with Holley and Wendell, Chan and Eden form a close-knit team bent on solving the unspecified crime. Wong unexpectedly returns after visiting a sick relative in San Francisco, but is dispatched by a stab to the side the moment he crosses the threshold onto Madden's property; he consequently becomes the novel's second and last murder victim (if we count Tony as the first).

The two deaths come in quick succession. Tony dies in chapter 6 while Wong is away in San Francisco. Wong arrives back in Eldorado in chapter 9 and has been murdered by the end of the chapter, around three pages after his reappearance. The proximity between the two deaths becomes a more significant thematic link via Chan's initial exclamation on the discovery of Wong's body, mysteriously stabbed in the side as he gets out of the car that drove him back to Madden's ranch: "Dead—like Tony" (127), Chan observes. Shortly afterwards, Chan develops his simile into a fuller hypothesis as he and Bob Eden grapple with the mystery:

> "murder of Louie just like death of parrot—one more dark deed covering up very black deed occurring here before we arrive of mysterious scene. Before parrot go, before Louie make unexpected exit, unknown person dies screaming unanswered cries for help. Who? Maybe in time we learn".
>
> "Then you think Louie was killed because he knew too much?"
>
> "Just like Tony, yes" (133).

Superficially, then, the resemblance between the two deaths operates through their common function in the novel's narrative. Wong and Tony are apparently bearers of information which requires that they are disposed of so that the underlying "black deed" can remain undiscovered. More importantly, the disposal of Wong and Tony suggests a wider disposability that the Chinese-American servant and the cosmopolitan parrot share in the text; neither of these deaths forms the central focus of the investigation, though the response of the novel's cast to these two deaths is telling.

Although Wong is universally liked and respected by the Eldorado community, his death is greeted in some quarters in starkly racist terms. When the local police officer Constable Brackett arrives on the scene to investigate the murder, Madden downplays the seriousness of events: "[F]ortunately no one was hurt. No white man, I mean. Just my old Chink, Louie Wong" (130). Any sense that Biggers is straightforwardly reproducing and endorsing the prejudices of his era is immediately complicated by the arrival of Chan in his disguise as the new cook: "Ah Kim had entered in time to hear this speech, and his eyes blazed for a moment as they rested on the callous face of the millionaire" (130). Moreover, we later discover that the racist Madden is not who he seems. It transpires that Madden is actually Delaney, a man who makes a dishonest living by impersonating tycoons. Working with the secretary Thorn and the good-for-nothing Maydorf brothers, Delaney was plotting to steal the pearls and to fleece Madden for whatever else he could get. The critique of racism emerges therefore both through the ferocity of Chan's gaze and in the attribution of the brutal dismissal of Wong's significance to one of the novel's principal villains. Wong's life in Delaney's view functions in Judith Butler's terms as "an ungrievable life": a "life that cannot be mourned because it has never lived, it has never counted as a life at all".[16] Importantly, and contrastingly, Wong's death is systematically highlighted as a cause of grief to the novel's other characters, even those with relatively minor roles. Holley is interrupted at his work in the newspaper office "dashing off poor old Louie's obit" (156); Doctor Whitcomb immediately concerns herself with "the murder of poor Louie Wong" (148). The novel carefully creates a discrepancy between Delaney's racism and the humanity of the rest of the novel's cast in a way that unsettles the hegemonic racial hierarchies that often formed a part of the Golden Age of detective fiction.

The heyday of detective fiction in the 1920s and 1930s involved a consistent entrenchment of the era's racial inequalities. Maureen T. Reddy notes of hard-boiled fiction (though the claim also works for other strains of the form in the period) that the "dominant consciousness [...] is indisputably a white consciousness".[17] Biggers can claim, therefore, a notable place in the genre's history. The first book in the Charlie Chan series, *The House Without a Key* (1925) can be considered, in Reddy's words, the first novel "to feature a detective of color".[18] This important development had significant ideological force. Biggers' created Chan as a specific counter move to the prevalence of the "yellow peril" narratives that was so central to the British author Sax Rohmer's Fu Manchu stories (1911–1959) and across the Atlantic in the work of a great many detective authors, including scions of the genre like Dashiell Hammett in "Dead Yellow Women" (1925), for example. William F. Wu identifies the trope of the menacing Asiatic figure as one that "dominates American fiction about Chinese Americans between 1850 and 1940".[19] It is worth noting that *The Chinese Parrot* was published just three years before Ronald Knox produced his

"Ten Commandments of Detective Fiction" including the instruction that "No Chinaman must figure in the story", an edict which both testifies to and critiques the formulaic introduction of malign Asian characters in so many detective novels.[20]

The setting of *The Chinese Parrot* in California has particular significance; debates about the rights of Chinese immigrants were especially important here given the extent of Chinese labour in the state's agricultural and construction sectors. The creation of Charlie Chan has a progressive, even radical force. On the author's untimely death, S. L. Yu wrote in *The Boston Traveller* that through Chan Biggers had blazed "a new trail of inter racial understanding".[21] As the Charlie Chan series grew in popularity, Biggers fought energetically against the reduction of his detective back into a racial stereotype in comic strips, films and on the covers of his books.[22] Certainly, the portrayal of Chan is not without elements of Orientalist essentialism (Chan repeatedly emphasises the psychic qualities of the Chinese, for example), but *The Chinese Parrot* derives much of its energy from its rebuttal of racist attitudes against Asian-Americans, particularly in the way it circulates around a self-conscious anti-racist joke. In order to get away with his disguise as Ah Kim, Chan is forced to play the part of the racist caricature in a way that grates against his sensibilities. He draws the line at one prominent aspect of racial stereotyping: "by the bones of my honourable ancestors, I will not say 'velly'" (the formulaic rendering of a Chinese pronunciation of "very", 68). The narrative pivots on the possibility of turning racism against itself; it is because he is able to pass as the surly Chinese cook that he is able to unravel the mystery. Admittedly, the novel's progressive position is undermined by the narrative side-lining of Wong's murder. Wong's death is a minor part of the narrative; it is not the crime that Chan and his associates are primarily concerned with. Nonetheless, the fact that Louie Wong is rendered grievable is a key part of the novel's conception as a (flawed) critique of the racist agenda of "yellow peril" fiction.

But who grieves for the parrot? Although various characters express a conventional regret for the bird's murder, it is only Wong that displays anything like an emotional response. When Holley breaks the news of his parrot's demise, Wong is devastated. In a pivotal scene, the narrative offers an extradiegetic reflection on Wong's grief in relation to stereotyped conceptions of Chinese identity:

> Anyone who believes the Chinese face is always expressionless should have seen Louie's then. A look of mingled pain and anger contorted it, and he at once burst into a flood of language that needed no translator. It was profane and terrifying (126).

Then, as Bob Eden drives back to the ranch (and towards Wong's own murder), the caretaker continues his lament in a passage that formulates the expression of grief through the aesthetics of the desert landscape:

> The moon had not yet risen; the stars, wan and far-off and unfriendly, were devoid of light. They climbed between the mountains, and that mammoth doorway led seemingly to a black and threatening inferno that Eden could sense but could not see. Down the rocky road and on to the sandy floor of the desert they crept along; out of the dark beside the way gleamed little yellow eyes, flashing hatefully for a moment, then vanishing forever. Like the ugly ghosts of trees that had died the Joshuas writhed in agony, casting deformed, appealing arms aloft. And constantly as they rode on, muttered the weird voice of the old Chinese on the back seat, mourning the passing of his friend, the death of Tony (126).

Moments later Wong is dead. Although the cause of death is literally a knife wound, the pace of the scene creates a strong figurative connection between Wong's emotions and his demise; his vocal mourning seems to make it impossible for him to survive. Grief slips into death. There is another slippage here too. The "weird voice of the old Chinese" evidently forms part of the scene's gothic ambience in a way that weaves Wong into the spectral landscape. The "little yellow eyes" gleaming from the dark, "flashing hatefully for a moment, then vanishing forever" suggest that Biggers' flight of purple prose carries a political point, as if it is the stereotypical image of the malign Chinaman that is being disposed of via the narration of Wong's grief. We might even go so far as to think that Wong's grief operates metonymically as an expression of the experience of Chinese labourers in the California desert. This was a landscape shaped by Chinese labour, particularly in the construction of the railroads from the late nineteenth century onwards (a history that is alluded to several times in the text). As the sound of Wong's grieving continues throughout the journey to the ranch, the appealing arms of the Joshua trees, writhing in agony, function as a counterpart to the grief for the parrot; the landscape embodies and widens Wong's dismay to gesture towards a wider Chinese-American experience in California.

The parrot therefore becomes part of a chain of grief that circulates through the narrative: Wong's lament for the bird is tangled up, albeit with a good deal of subtlety, with the novel's lament for the experience of the 20,000 Chinese-Americans who laboured in California. In this sense the bird is condemned to operate figuratively. The parrot points away from itself to other stories; that, of course, is its structural role as an appendage to the central human narrative. When the novel's mystery is solved, Tony returns into the characters' thoughts in a way that tellingly highlights his significance, as well as undercutting the narrative's apparent investment in its Chinese-American characters. As events are tied up, Chan asks this of Sally Jordan's son Victor: "[B]efore returning north, it is fitting that you place wreath of blossoms on grave of Tony, the Chinese parrot. Tony died, but he lived to splendid purpose. Before he passed, he saved the Phillimore pearls" (271). Since it is his facility of speech that constitutes his central contribution to the solving of the case, it seems in the end that the issue is perhaps not so much "can he suffer?" as "can he talk?" Moreover, what matters in the end, even to Chan, is the continuing circulation of capital (in the form of the pearls) through the novel's white characters. If grief for the death of the bird is a way of foregrounding Biggers' political agenda, it is also the point at which his political agenda collapses into the all-too-familiar foregrounding of white America.

The Logic of the Commodity in *The Case of the Perjured Parrot*

There are a number of close resemblances between *The Case of the Perjured Parrot* and *The Chinese Parrot*. The resolution of each novel pivots on mistaken identity. Just as Madden is not Madden, a revelation about Sabin's identity is part of the intrigue in *The Case of the Perjured Parrot*. There is a suspect secretary in each case: Thorn in Biggers's novel and Waid in Gardner's. The most obvious link between them (aside from the focus on parrots) is the central importance of millionaires. As with *The Chinese Parrot*, the intrigue of Gardner's novel largely circulates the acquisition, retention and misappropriation of fantastic wealth. The novel begins with the arrival of Charles W. Sabin at Mason's office to request assistance in the investigation into the murder of his father Fremont C. Sabin. As Mason takes up the case, he pieces together a complex marital situation. Charles' mother (Fremont's first wife) died some years ago, after which Fremont had fallen for the wiles of the opportunistic Helen Watkins Sabin who, by the time we join the action, he is in the process of divorcing. Much of the plot revolves around whether Helen and Fremont had been legally divorced by the time of his murder, and thus who should rightfully inherit Fremont's riches. Eventually, the murderer is discovered to be the avaricious Waid who had long been forging cheques to plunder his employer's millions and who had hatched a plan to implicate Sabin's wife. Just as Tony's significance in Biggers's novel resides in his service to the Phillimore pearls, so in Gardner's the parrots' significance comes from their role in allowing Mason to protect capital by solving a financially-motivated crime. This economic context is emphasised in *The Case of the Perjured Parrot* by the exchange of various birds which reveals parrots not just as creatures caught up in financial intrigue, but as themselves commodities who are bought, sold and mistaken for each other throughout the novel.

The exchange of birds in *The Case of the Perjured Parrot* is sufficiently complex to require further explanation. Fremont Sabin had long had a pet parrot called Casanova, a gift from his brother, Arthur. When Casanova starts making troubling remarks ("Helen … don't shoot", etc.), Arthur, something of a parrot fancier himself, suspects his brother is in danger and borrows Casanova "to study the parrot, trying to get a clue to the person who had been teaching him" these ominous phrases (190). In order to avoid raising suspicion, Fremont then buys another parrot to cover for Casanova. This parrot is discovered by his corpse and is presumed by the police to be Casanova. Charles is able to inform Mason, however, that this was not in fact his father's companion animal, but an avian impostor; Casanova could easily be recognised, we learn, by his distinctive missing claw.

Mason quickly sees the parrots as key to the case. His first port of call is a local pet store where it is confirmed that Sabin senior had recently purchased a parrot. The pet store owner also mentions one Helen Monteith, a librarian who had visited the shop to buy some parrot food and to get some advice on parrot care. Mason visits Monteith's home and discovers Casanova on the porch squawking "Put that gun down, Helen!" (47). When Mason catches up with Monteith we learn that she was recently married to Fremont Sabin believing him to be one George Wallman, a down-on-his-luck grocery clerk. Things look bad for Helen, particularly when it turns out she recently took a gun from a library display case. To make progress in the investigation, Mason decides he needs to have a closer look at Casanova. To do this without attracting attention, he buys another parrot from the pet store (this is now parrot number three) and switches him for Casanova. The new parrot is promptly killed, its head "snickanseed off … with a butcher knife" (109); the killer, we later discover, is Waid who endeavouring to cover his tracks by offing Casanova (it was he who taught the bird the ominous phrasing in a bid to frame Mrs Sabin), accidentally kills the wrong bird. Helen is arrested and at the coroner's inquest to determine the cause of Fremont's death it looks like her chances are up, especially when she quails when presented with the corpse of the dead parrot and her dismay is interpreted by the investigating police officer as a sign of guilt. Mason astonishes the court however by revealing Waid as the true culprit. Consequently, we end the novel with three parrots: one dead and two alive. Each of these birds functions in effect as a version of the same parrot. The original parrot is Casanova; the other two are impostors brought in to cover for Casanova while the lead parrot is examined for the secrets he appears to hold.

The novel's parrot swapping antics are premised on the existence of a ready supply of parrots. Whenever anyone needs a parrot, a parrot can be found. The novel's engagement with this context develops most strongly through the depiction of Mason's visit to the pet store. In terms of the novel's main drama, the scene is brief and incidental and the store's proprietor Arthur Gibbs is a decidedly minor character. Yet, as the source of the text's two dummy parrots, the shop underpins the narrative. Importantly, Mason's visit to the shop to interrogate Gibbs surrounds the trade in exotic birds with a subtly morbid atmosphere that contrasts with a depiction of wild birds in the previous chapter and inflects more dramatic scenes later. On first meeting the shopkeeper, Mason shakes "a bony, long-fingered hand which seemed completely lacking in initiative" (40). Shortly afterwards, as Gibbs reaches for a telephone directory, the narrative notes his "long listless fingers" (44). Gibbs' eyes, moreover, have "the colour of a faded blue shirt which has been left too long on the clothesline" (40). Here is someone who trades in living creatures and is himself drained of life and enthusiasm, a condition which suffuses the shopkeeper's work as he evinces considerable cynicism towards the motivations of parrot owners. Most crucially, part of Gibbs's task as a parrot trader is to "deliberately train them to swear" (41). Parrots do not acquire human language spontaneously, but only as the result of careful and laborious training. In Marx's terminology, it is Gibbs's labour that to a significant degree renders the natural object (the parrot) as the economic object (the pet).

As befits the world-weary atmosphere of the shop, Gibbs is scathing about his task. He notes that "you'll find people who think it's smart to have a parrot that cusses. Usually they get tired of them before they've had them for a long while" (40–41). Parrot-keeping in Gibbs' jaded world is the expression of a tawdry quest for a soon-exhausted sense of novelty. In this context, Mason and Street's visit to Sabin's cabin in the previous chapter comes into sharper focus. Here again we have talking birds, but not the potty-beaked parrots of Gibbs' store. Reflecting on the cabin's "beautiful setting", Mason is suddenly interrupted: "A blue jay resenting their intrusion, launched himself downward from the top of one of the pine trees, screeching his raucous, '*Thief… thief… thief*'" (21). The repetition of "thief" is a conventional representation of the blue jay's alarm call, but the semantic sense of the bird's cry also resonates into the ensuing pet shop scene as an expression of dismay at human interference with birdlife. The birds in Gibbs' store do not get to speak for themselves (all they can say in human terms is what Gibbs teaches them), but the free voice of the jays emphasises the novel's discomfort at the exotic bird trade.

The sense of unease at the commodification of parrots that the depiction of the pet store produces is developed through the plot's ready interchangeability of parrots. Parrots find their meaning in the plot through their circulation between owners. Rather than seeing birds as individuals with their own lifeworlds, the novel's cast see at the general level of fungible, mass-produced objects, a disposition that the plot subtly critiques. Gibbs reports a visitor to the shop (Fremont Sabin as it turns out) buying the first Casanova substitute and specifying the "breed and size and age of the bird" (42) in the understanding that one parrot of a certain species might easily be replaced by any roughly comparable other bird of that species (given the other restrictions of size and age). That the police presume that the parrot discovered at the scene of the crime is Casanova confirms the view of any parrot being much the same as any other parrot. Crucially, the investigation of the mystery works through the discovery that this logic is ultimately false. It is because of the singularity of Casanova's missing claw that the parrot-switching becomes apparent, which leads Mason to Monteith and, eventually, to the resolution of the narrative. Likewise, it is because Waid fails to recognise the individuality of the birds that he murders the wrong parrot and unwittingly helps to guide Mason towards the truth, as the lawyer explains in his closing summary of the case (182).

What is at issue, therefore, in Gardner's psittacine merry-go-round, and in the wider practice of the parrot trade, is the balance between the particularity of individual birds and the generality of the species.

Against this background, the climactic scene in which Helen Monteith is presented with the remains of the beheaded parrot carries significant force as an insistence on the ethical depth of bird life in contradistinction to the generic functioning of the parrot as commodity. At first as the district attorney Sprague asks his deputy to fetch the dead parrot, the bird appears as an enticing if gruesome spectacle for the gathered crowd:

> The deputy stepped outside long enough to pick up a bundle covered with cloth, then hurried down the aisle, past the rows of twisted-necked spectators, to deliver the bundle to Sprague.
>
> District Attorney Sprague dramatically whipped away the cloth. A gasp sounded from the spectators as they saw what the cloth had concealed—a bloodstained parrot cage, on the floor of which lay the stiff body of a dead parrot, its head completely severed.
>
> "That," the district attorney said dramatically, "is *your* handiwork, isn't it, Miss Monteith?"
>
> She swayed slightly in the witness chair. "I feel giddy," she said "...Please take that away ... The blood..." (159).

The moment's heightened atmosphere pivots on Sprague's flamboyant whipping away of the cloth in front of the "twisted-necked spectators". The parrot now becomes a different form of commodity: No mere pet, Casanova is the prop in a misguidedly triumphant performance. Sprague's energy contrasts sharply with Gibbs's listlessness, but the effect on the representation of the parrot is comparable. It is not the parrot as a sentient being that entices the crowd, but the thrill of the revelation which installs the parrot as an object of spectacle. Monteith's giddy reaction then changes the mood. Certainly, her response emphasises the scene's drama, but even if her plea to "take that away" confirms the bird's status as a spectacular object, her quavering attention to the blood hints at something different. As a conventional synecdoche for life, the image of the blood emphasises a recognition of the parrot's autonomous existence as part of Monteith's experience of the moment's trauma. Monteith quavers because she is faced with the bird's irreducible materiality as a vulnerable body in tension with the plot's alternative designation of the parrot as a unit in the circulation of commodities.

Shortly afterwards, Mason astonishes the crowd by producing another bird, this time the living form of Casanova complete with missing claw. As Monteith identifies Sabin's original companion, the pieces of the puzzle start to fall into place, Sprague's case falls apart and Waid quickly realises the game is up. Although the narrative is mainly concerned, as the detective form dictates, with establishing the facts of the case, Gardner provides a brief moment of closer attention to Casanova as Monteith prepares to make her identification. This parrot is a lively and rather quizzical presence:

> The parrot twisted its head first to one side, then the other, leered about him at the courtroom with twinkling, wicked little eyes; then, as Mason set the cage on the table, the bird hooked its beak on the cross-wires of the cage, and completely circled it, walking over the top, head downward, to return to the perch as though proud of the accomplishment (162).

After the spectators twisted their necks to catch a glimpse of Sprague's gruesome revelation, it is now the parrot's turn to twist its neck. The echo in Gardner's phrasing provides an ironic reversal that foregrounds the agential force of the bird. Casanova turns his bird gaze back on the spectators with a note of defiance. The wickedness of his eyes emphasises that he is no mere pet, but his own being with his own interiority and capacity for agency.

The scene moves on quickly to forms of engagement in which Casanova's view of things is suddenly less significant. As Mason looks to calm the bird with a casual "Nice Polly", we are returned to a generic figuring of the parrot. Casanova then repeats his refrain of "Helen … don't shoot" (162) and a discussion of the legitimacy of the parrot's apparent accusation of Monteith ensues with the result that the coroner rules that "a parrot can't be a witness, but the parrot *did* say something" (original emphasis, 162). The parrot is thus translated back into the kind of philosophical concern that occupied Descartes; the question once again is not so much whether the bird can suffer, but whether it can reason (enough for its articulations to be admissible as evidence). Gardner does not leave the parrot there, however. There is a last twist as it turns out that Helen Monteith's husband is not after all the murdered man. She had in fact married Fremont's reclusive and eccentric brother Arthur and is reunited with him at the novel's somewhat schmaltzy conclusion. In the novel's closing summary, Arthur reveals his part in the parrot plot. Particularly, he explains why he was led to substitute another parrot for

Casanova: “parrots vary, you know, and I knew Casanova” (190). The bird’s singularity, as opposed to its fungibility, is the final note of the parrot plot and a theme that gains particular resonance when we learn a little more about Arthur’s character.

Shortly after discovering he has inherited a large part of his brother’s fortune, Arthur asks his wife “should we take enough to open up a little grocery store, or shall we tell ‘em we don’t want any”. She confirms his view with familiar folksy wisdom: “Money doesn’t buy happiness” (191). Arthur’s rejection of capital and his sensitivity to the bird aligns animal ethics with a rejection of an overemphasis on economic achievement, as the novel concludes as a gentle anti-materialist fable. The novel’s closing scene sees Mason and Street counting their blessings and driving away through the moonlight while Mason reflects “That’s quite an idea, to go through life doing your best work and letting the man-made tokens of payment take care of themselves” (192). As the novel closes on the image of Street’s hand resting on Mason’s on the steering wheel, it is clear that the social order is largely restored: what could be more hegemonic than the triangle of man, woman and car? The novel’s parrot ethics is ultimately, as with Biggers’s novel, inconsistent: The narrative works towards an identification of ethical obligations to parrots, and to emphasise the being of parrots beyond their capacity to emulate human speech, but stops short of setting out any wider political implications of this apparently pro-animal position.

Conclusion

It can seem hard to take parrots seriously. Apart from anything, the world is full of parrot jokes.[23] There is an underlying whimsy to the idea of the parrot witness that creates laughter in the intrusion of animals into the resolutely anthropocentric domain of the law, as with the case of the stolen Dublin parrot. Both Biggers’ and Gardner’s novels certainly do more than take recourse to the tradition of the comedy parrot. In each of the texts, birds are meaningful presences whose lives and deaths can, at certain moments at any rate, exceed the generic function of their voices as clues and the philosophical speculations to which parrot speech is inevitably tied. Both *The Chinese Parrot* and *The Case of the Perjured Parrot* attend to the bird beyond the word and show genre and ethics straining against each other in the depiction of parrots, even if neither novel is able to entirely break free from the facility of speech as the overriding function of the textual parrot.

As we have seen, there are subtly differing narrative investments between the texts. While Biggers takes us back towards capital in the novel's final insistence that the purpose of Tony's life is to redeem the Phillimore pearls, Gardner's denouement takes us away from capital—albeit quite gently—in the final anti-materialist plot twist (though the closing image in the car edges back again towards capital). Ultimately, in both novels talking parrots are very closely associated with global capital. In this context Perry Mason's visit to the pet store might claim an even wider significance. It is a parrot's ability to articulate human sounds which makes them such desirable companions. Since parrots do not learn human language on their own, the Enlightenment's philosophical attraction to parrots is premised on the kind of commodification that Gibbs works at so wearily in his shop. Parrot talk is necessarily premised on violence. The figure of the parrot in detective fiction might, therefore, be taken to articulate in singularly evocative terms the collision between economic, narrative and philosophical modes of signifying animals. As Paul Carter trenchantly puts it, "parrots had to be enslaved to be known".[24] It is the trade in parrots which creates our understanding of them. If what we think we know of animals emerges in part out of an Enlightenment philosophical tradition in which parrots take a surprisingly central role, then what we know of other creatures comes from violence towards them. In both *The Chinese Parrot* and *The Case of the Perjured Parrot*, even the ethical commitment to other creatures remains bound to this violence through the perennially evocative idea of a talking bird. As Chan and Mason successfully solve their mysteries, there is a whole other set of crimes—against animals more broadly and parrots in particular—which remain unresolved. But we can at least in Biggers' and Gardner's novels find the beginnings of an investigation into the histories and mechanisms of this violence through the attention the birds come to demand.

Notes

1. "The Parrot Witness", 9–10.
2. "The Parrot Witness", 10.
3. *The Chinese Parrot*, 63. Subsequent references appear in parenthesis.
4. Gardner, *The Perjured Parrot*, 4. Subsequent references appear in parenthesis.

5. Boehrer, *Parrot Culture*, 147.
6. Boehrer, *Parrot Culture*, 147.
7. Descartes, *Discourse on Method*.
8. Descartes, *Discourse on Method*.
9. Locke, *An Essay Concerning Human Understanding*, 178.
10. Locke, *Human Understanding*, 180.
11. DiPiero. "Voltaire's Parrot," 341.
12. DiPierro, "Voltaire's Parrot," 342.
13. La Mettrie, *La Mettrie: Machine Man and Other Writings*, 12.
14. Bentham, *Morals and Legislation*, 311.
15. Wong. "An Interview with Alex."
16. Butler, *Frames of War*, 53.
17. Reddy, *Traces, Codes, and Clues*, 9.
18. Reddy, "Race and American Crime Fiction," 140.
19. Cited in Reddy, *Traces, Codes, and Clues*, 18.
20. Knox, "Introduction," np.
21. Cited in Westfall, "An Elegy to Charlie Chan," 52.
22. See Westfall, "An Elegy for Charlie Chan," 48.
23. Carter, *Parrot*, 180–182.
24. Carter, *Parrot*, 18.

Works Cited

Anon. "The Parrot Witness." *The Mirror*, 1. 1, Jan 4, 1845, 9–10.

Bentham, Jeremy. *An Introduction to the Principles of Morals and Legislation*. London: T. Payne and Son, 1789

Biggers, Earl Derr. *The Chinese Parrot*. Chicago: Academy Publishers, 2008.

Boehrer, Bruce Thomas. *Parrot Culture: Our 2500-Year-Long Fascination with the World's Most Talkative Bird*. Philadelphia: University of Pennsylvania Press, 2015.

Butler, Judith. *Frames of War: When is Life Grievable?* London: Verso, 2016.

Carter, Paul. *Parrot*. London: Reaktion books, 2006.

Descartes, René. *Discourse on Method*. https://www.gutenberg.org/files/59/59-h/59-h.htm

DiPiero, Thomas. "Voltaire's Parrot; or, How to Do Things with Birds." *Modern Language Quarterly* 70, no. 3 (2009): 341–362.

Gardner, Erle Stanley. *The Perjured Parrot*. Looe: House of Stratus, 2000.

Knox, Ronald Arbuthnott. "Introduction." In *The Best English Detective Stories of 1928*, edited by Ronald Arbuthnott Knox and Henry Harrington. New York: H. Liveright, 1929.

La Mettrie, Julien Jean Offray. *La Mettrie: Machine Man and Other Writings*. Cambridge: Cambridge University Press, 1996.

Locke, John. *An Essay Concerning Human Understanding*. London: Penguin, 1997.

Reddy, Maureen T. *Traces, Codes, and Clues: Reading Race in Crime Fiction*. New Brunswick: Rutgers University Press, 2003.

Reddy, Maureen T. "Race and American Crime Fiction." In *The Cambridge Companion to American Crime Fiction*, edited by Catherine Ross Nickerson, 135–147. Cambridge: Cambridge University Press, 2010.

Westfall, Mandy RK. "An elegy to Charlie Chan: Chang Apana, Earl Derr Biggers and Asian America." PhD diss., 2007.

Wong, Kate. "An Interview with Alex, the African Grey Parrot." *Scientific American*, September 12, 2007. https://blogs.scientificamerican.com/news-blog/an-interview-with-alex-the-african/

Ecology, Capability and Companion Species: Conflicting Ethics in Nevada Barr's *Blood Lure*

Karin Molander Danielsson

Nevada Barr's detective novels about park ranger Anna Pigeon are set in different US national parks that are integral to the series' plotlines and characters. In the first novel of the series, *The Track of the Cat* (1993), an illegal big game hunting operator murders a park ranger and covers up his crime as a deadly mountain lion attack, reinforcing the local farmers' hatred of lions and other big predators.[1] The ending identifies the human murderer and exonerates the lion, and Anna Pigeon is established as a protector of animals and supporter of environmental and animal preservation ethics, a stance that is repeated throughout the series. In the ninth novel, *Blood Lure* from 2002, the same plot device—a park animal implicated in a murder—is revisited but given a very different twist and an arguably problematic solution.[2] The primary concern of this chapter is the way

K. M. Danielsson (✉)
Sheffield, UK

Mälardalen University, Västerås, Sweden
e-mail: karin.molander.danielsson@mdu.se

R. Hawthorn, J. Miller (eds.), *Animals in Detective Fiction*, Palgrave Studies in Animals and Literature,
https://doi.org/10.1007/978-3-031-09241-1_6

the novel handles the animal as murderer motif and the three competing ethical positions that are signalled and developed in the text by the park setting, the representation of grizzly bears, and the relationships between human and non-human characters. According to Martha Nussbaum, textual ethics are disclosed in form as well as in content: "the selection of genre, formal structures, sentences, vocabulary, … all of this expresses a sense of life and of value, a sense of what matters and what does not".[3] In *Blood Lure*, that sense of *what matters* is compromised by an instability in the novel's generic and ethical affiliations. These affiliations are, on the one hand, the conventions of the detective genre and of the Anna Pigeon series, and, on the other, three different ethical schemas: *ecological ethics*, as defined and discussed by Donaldson and Kymlicka,[4] *capability ethics*, described by Nussbaum,[5] and Haraway's *companion species ethics*.[6]

Ecological ethics, according to Donaldson and Kymlicka, "elevate a particular view of what constitutes a healthy … ecosystem, and are willing to sacrifice individual animal lives in order to achieve this holistic vision".[7] In contrast, Nussbaum's capability ethics focus on the individual animal but in relation to the norm set by its species. The goal of the capability approach, Nussbaum writes, "is to take into account the rich plurality of activities that sentient beings need—all those that are required for a life with dignity … [and] see each living thing flourish *as the sort of thing it is*" (my emphasis).[8] Companion species ethics, finally, is Donna Haraway's conception of [human and animal] "species interdependence".[9] It is concerned with mutuality, how "diverse bodies and meanings coshape one another", and emphasizes interspecies, human–animal, "regard and respect".[10]

Blood Lure is set in Glacier National Park, one of the few remaining habitats of the North American grizzly bear. The plot follows Anna Pigeon and a team of researchers who are investigating the grizzly population, collecting DNA and mapping how individual bears move around the park. When a dead hiker is found, a bear attack is suspected at first but subsequently ruled out, since the face of the victim seems to have been sliced off with a knife. Nevertheless, in the last chapter a grizzly bear, albeit not of the park population, is pronounced guilty of the killing. The killer bear is a trained zoo animal called Balthazar, an enormous Alaskan grizzly. The zoo-owner had threatened to sell him to a game park, where he would be hunted for sport, which prompted the bear's teenaged trainer and companion, Geoffrey, to take Balthazar to Glacier Park, thinking he would be protected by park policy. Instead, Balthazar kills a woman tourist who is threatening him with bear spray, and the boy mutilates the corpse in order to cover up the bear's claw marks. In the end, after some intervention by Anna Pigeon who clearly admires "the magnificent beast" (332), the teenager and the bear are both placed in a training facility for movie animals, one as an apprentice, the other as an animal actor.

This ending is satisfactory to Anna, because it saves the life of a "wonderful bear" (332), and moreover, it saves what she recognizes as an enviable interspecies companionship between the bear and the boy (319). In the crime fiction framework, however, this solution seems inadequate in that it gives impunity to Geoffrey, a human who has acted deliberately in mutilating a dead woman, and also lets the killer bear live. From the perspective of ecological ethics, moreover, a killer bear that it is not put to death is problematic. It threatens public acceptance of wild bears and other predators in the park, and thus, in the long run, the ecosystem's balance of predators and prey. Finally, an ending which sends a grizzly bear into confinement also highlights the conflict between companion species ethics and capability species: it ensures that the boy and the bear can stay together, but disregards the bear's rights to a life where it can flourish *as a bear*.

These conflicts thus complicate both the novel's position in the detective genre and its stance with environmental and animal welfare ethics, long established by the Anna Pigeon series. I argue that the way *Blood Lure* circumvents these conflicts is by its strategy of splitting the image it presents of the grizzly bear into three different ones, with distinct discourses, associations and intertexts, thereby disassociating the wild bears in the park from the killer bear, and the killer bear from the one who is saved in the end. In what follows, I will first provide a background to the significance of detective genre conventions and national park policy, before I continue with a reading of the different images presented of the grizzly and the ethical frameworks suggested by how the bears are represented.

Conventions and Policies

Detective novels often follow one of three main traditions: classic whodunits, hardboiled PI-stories and police procedurals.[11] *Blood Lure* might be described as a hybrid text in the way it utilizes conventions from several subgenres. Classic "whodunit" conventions are employed when a violent death disrupts the status quo of a seemingly innocent community.[12] As Peter Messent argues, "The assumption underlying the detective stories featuring Holmes and Christie's Hercule Poirot—and, by extension, many other detective protagonists—is that the social body, representing the existing, and desired, good, is always the 'innocent' victim of the individual criminal act".[13] This innocent society must then strive to get back to normality by identifying the murderous individual and bringing him or her to justice.[14] In *Blood Lure*, the park, covering a limited area and constituting a specific, delicate ecosystem, can be seen as such a community, protected by the national park service's well-known commitment to "ecological principles" and policies based on scientific research.[15] Furthermore, Anna Pigeon's investigation leads to a classic "library scene" denouement, in a clearing in the woods, where the criminal is identified, and the mystery explained (321). Because the plot of *Blood Lure* aligns itself with the whodunit conventions in this manner, the anomalous ending, which names the killer but arguably does not bring him to justice, constitutes a challenge.

Although Anna is not a private eye, it is also relevant to read this series as part of the surge of new detective novels appearing in the 1980s and 1990s, with female protagonists and mainly hardboiled plots, characterized for instance by socio-political rather than private motives and cynical, flawed protagonists who not only investigate the murder but actively "interven[e]in events".[16] Walton and Jones, in their classic work on female hardboiled detectives, argue that these novels are characterized by a Foucauldian "reverse discourse … producing a critique of the formula by reproducing it with strategic differences", effective because of the "reader's recognition of differential intertextual relationships".[17] In Barr's series, that difference is partly created by Anna Pigeon's female reincarnation of the hardboiled hero—cynical, tough talking and often physically intervening—and partly by the detective's stance in relation to environmental and animal preservation ethics, established in the first novel. Bertens and D'Haen, similarly, recognize Anna's as "private and asocial", and "an independent investigator in all senses of the word", but nevertheless classify the Anna Pigeon novels as police procedurals, since the "protagonist works within, rather than outside of, the perimeters of institutional law enforcement".[18] Seeing that the national park service is inextricable from the settings and plots of the Anna Pigeon novels, this classification is also justifiable, regardless of her private eye characteristics.

In the first Pigeon novel, *Track of the Cat*, Anna Pigeon is a thirty-nine-year-old, wine-drinking misanthrope: "[A]lone was safe. A woman alone would live the longest" (15). In *Blood Lure* we meet a slightly older and more mellow Anna "somewhere in that fertile valley of middle age between forty-five and fifty-five" (4), who drinks tea (54) and enjoys the company of her colleagues (41), though her affinity with animals is unchanged and perhaps even enhanced by the iterations of national park settings, and by Anna's reflections about her "connection with the wild beasts" (3). This character development and the sustained environmental interest also position the Anna Pigeon novels alongside many other series with dynamic detectives and informed special interests, launched in the latter half of the twentieth century.[19] Like the authors of many of those series, Barr is able to draw upon her own work experience—she used to be employed as a park ranger—which may account for the fact that her "evocations of nature are simply breathtaking and all the more convincing because they are offered in a spirit of loving realism", as Bertens and D'Haen put it.[20] This claim for verisimilitude extends to other aspects than descriptions of nature. Each book is set in an actual park, they all include verifiable particulars of park geography, history, policy and management, and data about the typical park species. Reflecting the influential Leopold report of 1963, which insisted that park management used "the utmost in ... ecologic sensitivity, and be based on scientific research, [21] several Pigeon novels (e.g. *Blood Lure* and *Winter Study*) include authentic, named, science projects and scientists.[22] The insistence on authenticity in setting and plot creates an expectation of similar verisimilitude, or at least consistency, in the series' ethical position, but as I will show, this is a problematic aspect of *Blood Lure*.

Glacier Park in Montana accommodates a substantial grizzly bear population, the signature species of *Blood Lure*. Bears have long been tourist magnets in the national parks. Richard West Sellars claims that in the early decades of park history, "the Service sought to bring bears and people together, particularly with the bear shows at garbage dumps and by allowing roadside feeding, ... [but it also] removed 'problem' bears that threatened visitors or their possessions".[23] The number of such problem bears was not unsubstantial. Jeanne Clarke reports that in the years 1949–1963, "a total of 103 black bears and 15 Grizzlies were destroyed by park rangers" in Glacier Park, due to such conflicts. [24]

Despite many incidents, nobody was killed by a grizzly in Glacier until August of 1967, when two young women were killed the same night, in different locations.[25] During the years after these events, the park intensified its efforts to keep bears and tourists apart, for instance by imposing restrictions on the handling of food and garbage, and, in a move towards management based on research, by developing a bear-sighting database used to determine where tourists were to be allowed.[26] Even so, between the years of 1968 and 2017 eleven people were killed by grizzly bears in or near Glacier Park.[27] The plot of *Blood Lure*, in which two people are attacked and one person is killed, specifically reflects and refers to the 1967 incidents, adding to the novel's verisimilitude.

Despite other changes in park policy, bears that kill people in national parks are still relocated or put down. The Bear Management Guidelines for Glacier Park from 2001, near the time depicted in the novel, distinguishes between a bear that has been *habituated* to people, that is, "is tolerant of human presence, has become accustomed to frequenting … campgrounds, etc.", and a bear that has been *conditioned*, namely has "sought and obtained non-natural foods, destroyed property, torn into tents or backpacks, [or] displayed aggressive (non-defensive) behaviour toward humans".[28] Habituated bears "may be released on site with behaviour modification" (not specified), "or relocated", whereas conditioned bears will be "removed from the Park by relocation, or being [sic] destroyed".[29] That such guidelines exist is made clear in *Blood Lure*, as is the fact that identification of rogue bears is now possible through DNA matching, thanks to a research project aimed at estimating "grizzly bear population size and density" by means of DNA sampling[30] which was carried out in Glacier in 1998 and 2000, and which plays a significant part in the plot of *Blood Lure*. A biologist in the novel explains:

> We solved a bear murder case three years ago. Got DNA samples from the poop [on the site of the killing] and lo and behold, they matched up with bear samples we'd taken the year before from another bear/human interface. So we knew we had the right bears and weren't just killing them to make the victim's family happy (58).

An actual bear who, like the killer bear in *Blood Lure*, attacks a campsite, ravages human property and finally kills a person, would most likely be shot, or at the very least placed in a zoo. The fact that this novel's killer bear escapes both death and prison, therefore conflicts with the idea that the novel offers a largely authentic view of Glacier Park. It also largely disregards the ethical frameworks suggested by the text. In what follows, I will elaborate on these ethical perspectives, and show how the novel circumnavigates these conflicts by representing the grizzly as three quite different creatures, associated with their own situations, intertexts and ethics.

Ecological Ethics and the Image of the Species

The first image of the grizzly is that of the grizzly *species*, an indispensable part of the ecosystem, and an object for ecological study. The 1963 Leopold report on US national parks recommended that the National Park Service "recognize the enormous complexity of ecologic communities", and stated that "Management based on scientific research is … not only desirable but often essential", all of which is reflected in the novel.[31] The novel's second sentence establishes the connection between science and bears, introducing: "Joan Rand, the biologist overseeing Glacier's groundbreaking bear DNA project" (1). A couple of pages later Anna is seen studying a picture of a bear. Although Anna has some scientific training, and is assisting in a science project, she is not a scientist. Therefore, her description of the picture signals scientific association, rather than precision:

> [S]he studied the color photocopies of *Ursus horribilis* thumbtacked to a long bulletin board situated over the conference table: the muscular hump over the shoulders developed, it was thought, to aid in the main function of the four-inch claws—digging. Fur was brown, tipped or grizzled with silver, earning the bear its name. Ears were rounded, plump, teddy-bear ears; teeth less sanguine, the canines an inch or so in length, well suited to their feeding habits. Grizzly bears ate carrion, plants, ground squirrels, insects, and sometimes, people (3).

The teddy-bear ears provide a little comic relief in an otherwise serious paragraph. The Latin name of the species is the first feature signifying the objectivity and formality of scientific writing. *Ursus arctos* is the Latin name of the brown bear species, occurring in Eurasia and America alike, while *Ursus arctos horribilis* denotes the subspecies, the grizzly, found in continental North America. Here Anna uses the shorter form *Ursus horribilis*, with the thrilling, yet scientific sounding, Latin word *horribilis*. The rest of the description includes features of science writing such as the passive voice: "developed, it was thought", an emphasis on function: "to aid in the main function of the four-inch claws—digging", and an economical style of description: "teeth less sanguine, the canines an inch or so in length, well suited to their feeding habits". The word sanguine is used here in its metaphorical sense, but obviously chosen because of the Latin origin referring to blood, reminding us of the title, *Blood Lure*. These stylistic features emphasize the fact that Anna here associates herself with the researchers, but, like the last sentence's list of grizzly foods, they simultaneously point to something more chilling, the genre's promise of death: "Grizzly bears ate carrion, plants, ground squirrels, insects, and sometimes, people" (3). The image conveyed is of the grizzly as a species, a group of animals, whose anatomy, habitat and sometimes frightening eating habits are well known to science.

The research carried out in the novel reflects a joint U.S. Geological Survey and National Park Service project led by Kate Kendall and published in 2008 as "Grizzly Bear Density in Glacier National Park, Montana", which thus creates an intertextual relationship between the novel and a research article. There is no doubt as to which the project is; the name and affiliation of the researcher is given in the novel (6) and Kendall is mentioned in the Acknowledgments as having "answered countless questions" (n. pag.). There is more than one page of novel text describing how the team go about their work, naming the trails and areas they will cover, detailing what the work entails: "going into the old traps, collecting the hair, dismantling the traps, and setting them up in new locations" (6). In the article, the method is described in very similar terms:

> We placed one hair trap in each cell for 14 days, after which we collected hair. ... We placed each hair sample in a uniquely numbered paper envelope and passed a flame under the barbs to remove any trace of hair. We then dismantled traps and moved them to another site within each cell.[32]

Note the occurrence of the collocations *collecting/collected hair* and *dismantling/dismantled traps* in both texts. One objective expressed in Kendall's article is: to "estimate grizzly bear population size and density" (1694). A similar objective is represented in *Blood Lure* as: "it was hoped an accurate census of bears could be established, as well as population trends, travel routes and patterns" (7). The correspondences strengthen the claim for authenticity, and help establish an uncritical position toward the project as well as toward the ecological ethos, to which this image of the grizzly is connected.

Ecological ethics, or what Donaldson and Kymlicka, in *Zoopolis*, call the ecological approach, is a framework that "focuses on the health of ecosystems, of which animals are a vital component, rather than on the fates of individual animals themselves".[33] For Donaldson and Kymlicka, this kind of ethics is insufficient and ineffective in the protection of animals' rights, because human interests and human views will always prevail regarding which ecosystems are worth preserving, and which individual animals may have to be sacrificed for the good of the system.[34] Although the word ecology is not used in the novel, *Blood Lure* reflects the National Park Service's ecological ethics in descriptions of the DNA project and of park policy, for instance regarding the vegetation which is allowed to grow freely, with old trees falling where they grew (18), forest fires, which are allowed to burn out (22), and in discussions of the importance of global environmental policy: "The [larvae of the army cutworm] moths are a major source of protein for the grizzlies. ... See? Global. Spray wheat in Minnesota, starve a grizzly in the Rockies" (216). The pitfalls of the ecological approach are also seen, however, for instance in discussions of so-called bear incident management: "[The biologist] looked worried. One of her four-hundred-pound charges had misbehaved. The concern wasn't misplaced, considering what penalties humankind often extracted from other species for even the slightest infractions" (35). In other words, the grizzly species' existence in the park is contingent on its good behaviour around tourists. Here, in association with the ecological ethics, emerges an image of the grizzly as a part of a system, closely managed and supervised by scientists and the NPS.

Capability Ethics and the Image of Fear

The scientific discourse of the novel also invites another ethos, *capability ethics*, in that the scientific, objective representation of bear behavior establishes what the grizzly species is capable of in the wild. The capability approach to animal rights, as discussed by, among others, Martha Nussbaum, is significant for animal welfare, especially for animals confined in zoos, because it directs attention to the individual animal's opportunity to exercise the capabilities of its species. Nussbaum maintains that "the species norm ... tells us what the appropriate benchmark is for judging whether a given creature has decent opportunities for flourishing".[35] Thus, depriving an animal who "characteristically live a life full of movement" of that opportunity—for instance by keeping a bear in captivity—would mean to "give [it] a distorted and impoverished existence".[36] In *Blood Lure*, this ethos is called up by the researchers and park rangers who discuss how the wild bears in the park roam across very large areas (35), use their claws to dig deep for glacier lily bulbs (23) and ground-living rodents and insects (216), scratch on trees (19), forage for berries in the autumn (20), breed and raise cubs (17), all aspects of a grizzly's capabilities. Anna has reason to contemplate bear powers when she ventures into one of the distant areas where the grizzlies dig for cutworms: “She spotted five of the oval-shaped excavations … and was astounded once more at the sheer physical power of the grizzlies. In places the digging went down a foot or more, and the volume of rock moved was in the tons” (226). This and many other references to the determination and aptitude of the grizzly, clearly signal capability ethics.

Capability ethics is also connected with another image of the grizzly, which is necessary to create the suspense and murder conventional of the detective genre. The species that expertly forages for insects and huckleberries does sometimes also “eat people”, as Anna’s early reflection points out. In the course of the novel, the reader learns about the threat bears pose to tourists who do not follow the rules or are careless (13); for instance, grizzlies are capable of attacking camps where food or garbage is left unattended (58), and of maiming or killing people who threaten a sow’s cubs (17). The way *Blood Lure* succeeds in capitalizing on this dark side of the species, without unnecessarily incriminating the park bears, is by introducing another bear image, an image of horror, with its own intertextual connections.

This image emerges as soon as the expedition sets out. In this part of the novel, the reader follows Anna, the biologist Joan, and the young volunteer Rory, as they hike into grizzly country of the park, carrying glass bottles of strong-smelling liquid bear lure, in order to relocate and refill the hair traps. Fear appears first in the young man who, nervous about the glass bottle operation, imagines that he has got some of the smelly liquid on his hands and might attract a bear: “Rory’s face was tight and young with fear. His eyes had gone too wide” (38). Anna, while knowing that his fear is unfounded, helps him wash the non-existent smell off his hands: “Fear was a killer. Anna had seen people die of it, when their wounds weren’t anywhere near mortal. Rory wasn’t in that kind of trouble, but fear distracted. That in itself was a danger in off-trail travel” (39). The smell might be imaginary, but the danger of attracting a bear is not. From this point, we know that there is something to be afraid of, not just for the tourists, but for the main characters as well. The new image of the grizzly as an object of fear, resonant of horror and wilderness adventure genres, is hereby established.

In the night, Anna is wakened by the sound of a bear brushing against the nylon of the tent, a noise described in vaguely supernatural terms as a “soft exhalation, the sigh of the wind or a ghostly child” (42). Her normal composure around wild animals dissipates, as a big animal circles around their tent:

> With each circuit, Anna's Disney-born sense of oneness with her fellows of the tooth and fur faded. It was replaced by the lurid pen and ink illustrations she remembered from a sensationalized account of two women killed when she was in college, both dragged from their tents, mauled, killed and fed on in *Night of the Grizzlies* (42–43).

The intertext so deliberately suggested here, Jack Olsen's bestseller from 1969,[37] relates the two August 1967 human–bear encounters in which two women were killed by two different grizzlies during the same night, one in the Granite park campground, the other at Trout Lake, several miles away across very rugged terrain. Much of Olsen's very detailed account of these events and what led up to them, evokes ecological ethics, in that it focuses on how, in this earlier period of national park history, park employees' mismanagement and neglect of park regulations led to bears and humans repeatedly coming into close contact. The general impression conveyed by Olsen is that of the bears as victims of circumstance, rather than as intentional killers. However, Olsen, who had a long career as a true crime author, approaches the attacks themselves in very detailed, gruesome accounts and those are the ones with which *Blood Lure* enters into an explicit intertextual relationship. Olsen describes the first attack, upon Roy Ducat and Julie Helgeson in the Granite park campground like this:

> He had been sleeping soundly when … a single blow from a huge paw knocked both of them five feet away on the ground, … Roy opened his eyes long enough to make out the shadow of a bear standing on all fours above the helpless girl and tearing at her body…. Now Roy could hear bones crunching and Julie screaming out: "It hurts!" and "Someone help us!" He realized that the girl's outcries were receding down the hill and he thought that the bear must be carrying her off.[38]

The darkness of the scene creates a good part of the fear and confusion, emphasizing the surviving witness's lack of vision and agency. The single blow comes out of nowhere and all he can make out are shadows and crunching sounds. In *Blood Lure* a similar scene takes place: "[Anna's] senses stretched: blind eyes trying to see through two layers of tenting … A growl broke the night above them, and both women screamed. The growling increased in volume and moved down the length of the tent" (43). Here we recognize the failure to get a visual overview, and the reliance on hearing, from Olsen's account, but also the women's realization that this bear is capable of harming them.

The second 1967 attack, on Michele Koons at Trout Lake, was witnessed by a fellow camper, Paul Dunn, who had managed to climb a tree, and is related by Olsen thus:

> "Michele!" Paul shouted. "Get out of your bag! Run and climb a tree!
> "I can't" the girl shouted. "He's got the zipper!" ... "He's got my arm... My arm is gone! Oh, my God, I'm dead!"
> Paul Dunn saw the bear lift the sleeping bag in its mouth and drag it out of the circle of fire and up the hillside into the darkness".[39]

The fact that the girl might have escaped, had she not been trapped in her sleeping bag, is a detail that has been picked up by several accounts of the events.[40] This image, of the girl in her sleeping bag, dragged into the darkness, echoes horror movies in which monsters like King Kong carry off human victims, but also points to another frightening characteristic of grizzly bears, their capability of dragging a body some distance. In *Blood Lure*, Anna panics at being trapped in the collapsed tent (like Michele Koons in her sleeping bag), and when the two women finally get out and realize that Rory, the young volunteer, is missing, they experience a similar loss of agency to Roy and Paul in Olsen's account: "'The bear must have dragged him out, taken him into the woods,' Joan said. ... 'Do bears do that? [Anna] had to ask, *Night of the Grizzlies* notwithstanding" (49–50). In this problematic passage, Anna, the voice of reason, is asking her question of a biologist, but answers it herself by the second reference to Olsen as the uncomfortable truth hits her: Yes, this is also something that bears do; something that grizzlies are capable of.

Despite the danger, Anna sets out to search for any signs of Rory. Alone in the darkness she experiences a new wave of paralyzing fear which leads to the suggestion that this might not, after all, be a regular bear, that this bear is somehow different from the other bears of the park:

> The circling and circling, the sudden rage, the fury of the violence, the abrupt cessation, as if a malevolent plan were behind it.
> Walt Disney lied. It was the Brothers Grimm who had the right idea: witches baked little girls, stepmothers poisoned them, bears ate them.

"Get a grip" she whispered, dizzy with the nightmare she'd just dreamed. A bear's a bear's a bear (51).

The word malevolent, the suggested plan, and the cluster of evil fairy tale images contradict Anna's last statement, however, and allow an image of a supernatural, cruel, Brothers Grimm type of animal to temporarily darken the tone of the novel. The individual actions of this bear—circling, growling, trashing equipment—are fully within the range of grizzly capabilities, and thus presumably as worthy of respect as the bears' physical power when digging for worms. This is clearly not the case when its powers threaten the protagonists, however. Instead, this particular bear is anthropomorphized by the suggestion of evil design, and thereby disassociated from the regular species whose normal impetus for violence—hunger, fear, defence of cubs—Anna by now is familiar with. By introducing this new image of an individual, malevolent grizzly, the text summons deadly violence and suspense without sullying the reputation of the park's general grizzly population. As it turns out, there is indeed design behind this bear's activities, but it is the design of a human brain.

Companion Species Ethics and the Image of Brother Bear

The young man who disappears from Joan's and Anna's camp is eventually found alive and well (84), but not before a woman tourist is found dead with a broken neck and a mutilated face (70). The rogue bear is of course suspected by the ranger who finds her, but closer inspection shows that the woman's face has been sliced off with a knife, which leads to the conclusion that the killer must be human (70). Investigations continue, and toward the very end of the novel, Anna and Joan find themselves ambushed in the woods by a male suspect called McCaskil, who threatens them with a gun (312). They are miraculously saved when into the clearing walks a teenage boy accompanied by an enormous, beautiful grizzly bear, who appears to be a tame companion to the boy.

When the bear appears, we see him from Anna's point of view, as their mythical, benign saviour and protector: "Out of the wood padded the great grizzly, beside him the ... boy. ... The spinning effervescence of a fairy tale snatched up Anna's brain. This bear was with them, of them, glittering gold protector of babes lost in the woods" (318).

This, the third prominent image of the grizzly in *Blood Lure* draws on several different manifestations of human–bear relations. This bear belongs to a human, but also to human–bear culture.[41] In this image can be seen traces of bear myths: bear gods, bears fostering half-bear children, and bear–human metamorphoses, but also of fairy tales like Goldilocks, and stories about toy bear companions, such as Paddington and Winnie-the-Pooh. Another view is offered by McCaskil, the then murder suspect, who turns out to be the owner of a roadside zoo from which this bear has been stolen by the young boy, its trainer Geoffrey. To McCaskil, the bear is simply frightening: "'Keep that goddamn bear off me!' McCaskil cried, his voiced ragged from yelling. 'Balthazar doesn't like him,' Geoffrey said. 'When we were little he used to tease us something awful'" (318). McCaskil's fear is thus immediately neutralized by Geoffrey's emphasis on companionship, which Anna automatically accepts. She merges this with her own, fairy-tale one: "We. The boy and the great bear had grown up together. Staggered by the unreality of the scene, Anna found herself wondering if they were brothers" (319). Instead of letting the realization that Balthazar is a former zoo-inmate prompt a reflection on animal ethics and the problems associated with keeping wild animal in zoos, as might be expected of an animal lover, Anna's admiration is allowed to obscure her sense of animal welfare. Geoffrey's possible complicity in keeping and training a wild animal is also hidden by Anna's awe of his companionship with it, a manifestation of her own dreams.

The species interdependence demonstrated by Geoffrey and Balthazar suggests a third ethos: *companion species* ethics, as defined by Donna Haraway. Companion *species*, says Haraway, is a "less shapely and more rambunctious"[42] category than mere companion animals, "a not-humanism in which species of all sorts [that is, not only pets] are in question. … The ethical regard … can be experienced across many sorts of species differences. The lovely part is that we can know only by looking and by looking back".[43] Companion species ethics thus stresses mutuality, a looking back at and a responding to each other that resonate with how the bear and the boy are represented in *Blood Lure*. Like Donna Haraway and her agility dog Cayenne "training each other in acts of communication",[44] the boy and the bear have developed, not only a way of communicating which lets the bear perform tricks like dancing, juggling, or standing on his hind feet and roaring on command (324–5), but an emotional "co-shaping".[45] Anna reflects: "Whatever [Geoffrey] felt floated to the surface where it could be easily seen by anyone with eyes. Perhaps growing up brother to a bear had denied him humanity's greatest defensive weapon: the lie" (326).

In the denouement, Balthazar's and Geoffrey's presence in the park, and their role in the death of the woman, are explained. In order to protect Balthazar from being sold to a hunting reserve, Geoffrey has stolen him from McCaskil, and is now trying to release him into Glacier where he would, Geoffrey assumes, be protected. The plan misfires, because Balthazar, a zoo bear, is unfit for a life in the wild. Geoffrey, his companion, recognizes the bear's fear of other animals, his inability to find food, and how easily he is spotted because of his size. In an effort to frighten away the research party, Geoffrey stages the horror movie performance by letting Balthazar act like a wild bear: growling, juggling with equipment and trashing the camp, on cue. Later, the bear is surprised by a woman tourist, wearing a coat belonging to and therefore smelling like his violent owner, McCaskil. When she threatens him with bear spray, Balthazar strikes her, and breaks her neck. Geoffrey, in an effort to conceal the claw marks and protect Balthazar, then slices the skin off her face (328).

Even with this knowledge, Anna's admiration for Balthazar is unlimited, and unanalytical: "She wanted to be near Balthazar. She found unending delight in the play of sun and shadow over his fur, the lumbering grace of his walk, the sharp accents his claws made on his tracks in the dust" (331). She acknowledges the unreality she experiences in his presence: "Balthazar … sat on huge haunches, ancient eyes watching like a primitive god" and the scene "between the trees took on a mad-tea-party feel" (320). Simultaneously, however, she clearly envies Geoffrey's companionship with the bear: "more than anything she wanted to touch him, play with him, listen to the stories he might tell" (319). No doubt Anna's reaction is meant as a confirmation of her love for animals, but her thoughts here are disturbing because they ignore the fact that the bear has killed a woman, and the fact that his life in captivity has possibly been instrumental in making him dangerous.

When Species Meet does not address the question of how *companionship* affects wild animals' species-specific needs, in other words, whether training—for example, to growl on cue— can stand in for the opportunity to exercise other *capabilities* bears show in the wild, such as roaming, digging, or mating. Some ethologists do argue, however, that the training of zoo animals constitutes enrichment, among other reasons because reward-based training can "trigger covert species specific behavior" and emotions.[46] Balthazar has had his immediate needs taken care of, he has been protected from the violent zoo owner, and has been trained to play, do tricks and perform in front of audiences. However, the question of whether the bear's situation brings opportunities for *flourishing* that alter or reduce his needs to exercise the other capabilities of his species, is not raised in *Blood Lure* at all. As soon as Balthazar is identified as a tame bear, his companion species status, his exceptional appearance, and his fairy-tale qualities preclude any comparison to wild grizzlies, or to capability ethics. In that sleight of hand, and especially perhaps, in the invocation of myth and fairy tale, an opportunity for an ethical discussion based on real animal welfare is lost.

Discussion and Conclusion

At the end, the question is not, "whodunit?" but "what will happen to Balthazar?" When asked this question by Geoffrey, Anna optimistically answers "Nothing bad" (329). Given her attachment to the bear, an ending involving his death seems unlikely, even though it would be realistic. Balthazar is after all what the Bear Management Guidelines would call *conditioned* to humans, unafraid, dependent on humans for food, and inherently violent. He is a prime example of what close contact between humans and bears can lead to, and what park policy expressly wants to avoid. Detective fiction conventions, and the ethics of ecology, capability and companion species, which have informed different parts of the novel, come here into conflict. Without Anna's investment in this individual bear and his trainer, death, or some kind of zoo-prison, is the ending we would expect, because detective novel conventions ask for the guilty party to be punished, so the society can return to its state of innocence. Granted, the question of whether an animal can be considered guilty and therefore sentenced, complicates this point. Animals have no legal standing; they can neither be taken to court, nor bring lawsuits for suffered injuries,[47] and are seldom granted agency in situations involving humans.

In the ecological, national park context, however, there *is* an established routine for dealing with killer bears: Investigations are made to make sure that the right bear is identified, trapped, and removed or euthanized,[48] because, as seen in *Blood Lure* (58) and in the "Bear Management Guidelines", individual killer bears must be destroyed in order to protect the species and the park. Complicating this issue is the fact that Balthazar is not a park native; he has never belonged to the park ecosystem, and this affects the applicability of the established practice. An Alaska grizzly would not be included among the species native to Glacier, and the Leopold report states that "management be limited to native plants and animals" (5). Moreover, a tame bear would be unable to fend for himself among the existing grizzly population. There is also the fact that this bear is the property of McCaskil, but since this man is a criminal awaiting charges he is in no position to object to whatever happens to his bear (325). Nevertheless, the bear has a monetary value to which can also be added the companion value Balthazar has to Geoffrey, who has grown up with the bear and who has trained him for their joint performances. Companion species ethics thus cannot be ruled out either, and would speak against destroying Balthazar,

There is yet another point speaking against punishing the bear altogether: Geoffrey makes a plea for mitigating circumstances. He points to the fact that the woman was wielding bear spray, and was smelling like McCaskil: "When he smells [McCaskil] he knows that something bad is happening and goes back to bear rules to save himself", (324) Geoffrey explains, construing the killing as an act of self-defence, which has been recognized in at least one deadly bear–human interface in Glacier.[49] This plea is double-edged however, in that it also shows the tame bear (possibly damaged by its life in captivity) reverting to a wild, pre-companion state, which reduces Balthazar's companion species status, and complicates Geoffrey's role in the killing.

Furthermore, although Geoffrey's plea effectively invites capability ethics, by indicating that this grizzly is a wild animal with all the abilities, reactions and strengths of its wild relatives, Balthazar's rights, as a member of the bear species, to opportunities for flourishing (Nussbaum) are left without any comment or consideration in the novel. While *Blood Lure* implies that Balthazar's thwarted upbringing in a zoo is what makes him different from the wild bears—more likely to get into close contact with humans and therefore more likely to be dangerous—the implications are never made explicit and no criticism of the practice of keeping bears confined is ever voiced. This becomes clear when the solution to the problem—what to do about Balthazar—is to send him back into captivity, albeit a slightly different one than the zoo. Anna actually contacts a number of zoos to ask them to take care of the bear, but the solution, presented as "great news" (333), is a training facility for animals used in the movie industry, which is also willing to take Geoffrey as an apprentice (333). This ensures that boy and bear will be able to stay together, and in fact recognizes Geoffrey's claim to his companion, but it does not consider any rights or needs of Balthazar's. While it might be argued that training could constitute opportunities for flourishing,[50] that possibility is in fact never mentioned in the novel, and Balthazar is never asked for his opinion. From the point when we learn that Geoffrey and the bear are together, Geoffrey in effect interprets for Balthazar; we see few actions by the bear that are not glossed by his human companion (327). Seen in this light, Balthazar's companionship status, which is what saves his life in the end, also underlines his dependency on Geoffrey, thwarts his agency, and serves to distinguish this bear manifestation from that of the wild species and that of the horrifying rogue.

Blood Lure is part of a detective series set in national parks, and as such includes a strong environmental and non-human presence, as well as a certain insistence upon accuracy of verifiable data, for example regarding scientific projects and park management. The genre conventions, the ecological ethics that accompany the national park setting, and the series' self-imposed call for verisimilitude, have however been complicated in this novel by a plot which includes a killer bear in a national park setting, but excludes the genre's most conventional, and the setting's most realistic, ending to such an event. This chapter has shown how the plot has necessitated a splitting of the novel's representation of the grizzly into three quite different images. The form in which these images are presented, and the accompanying intertexts, are indicative of three ethical frameworks; however, these frameworks have also been shown to be incompatible, which has had detrimental effects on the generic and ethical foundations of this novel. In the final analysis, what the textual ethics signal, what matters, as Nussbaum puts it, in *Blood Lure*, is ecological ethics: ecosystem before individual. However, with the novel's fairy-tale ending, even that ethical position is destabilized, as are *Blood Lure's* affiliations with genre conventions and verisimilitude. Furthermore, the novel's reluctance to initiate, much less sustain, a discussion of grizzly bear welfare, severely jeopardizes the novel's ethical relation to non-human animals.

Notes

1. Barr, *The Track of the Cat.*
2. Barr, *Blood Lure.*
3. Nussbaum, *Love's Knowledge*, 5.
4. Donaldson and Kymlicka, *Zoopolis.*
5. Nussbaum, "The Moral Status of Animals."
6. Haraway, *When Species Meet.*

7. Donaldson and Kymlicka, *Zoopolis*, 4.
8. Nussbaum, "Moral Status", 33.
9. Haraway, *When Species Meet,* 19
10. Haraway, *When Species Meet,* 4, 19.
11. Several other subgenres exist, of course, but this is a common division, used for instance in Bertens' and D'Haen's "Introduction" to *Contemporary American Crime Fiction* (Basingstoke: Palgrave, 2001).
12. W.H Auden, whose essay "The Guilty Vicarage" probably is the origin of the idea of the innocent society, actually points out that initially, that innocence is false, because there is a murderer, a fallen member of the society. When the murderer is identified "innocence is restored" ("The Guilty Vicarage", 408).
13. Messent, *Crime Fiction Handbook*, 13.
14. See also Knight, *Form and Ideology in Crime Fiction*, and Porter, *The Pursuit of Crime* for discussions of that innocent community and the justice imperative, both conventions which later detective subgenres have downplayed or departed from.
15. Sellars, *Preserving Nature, 3.*
16. Horsley. *Twentieth Century Crime Fiction*), 68, 75, 77.
17. Walton and Jones, *Detective Agency*, 92–93.
18. Bertens and D'Haen, *Contemporary American Crime Fiction*, 89, 77
19. Danielsson, *The Dynamic Detective.* Examples include Joan Smith's series about Loretta Lawson, Tony Hillerman's series about Joe Leaphorn and Jim Chee, and Barbara Neely's series about Blanche White.
20. Bertens and D'Haen, *Contemporary American Crime Fiction*, 86
21. Leopold, "Widlife Management Report", 4, 6.
22. Barr, *Winter Study.*
23. Sellars, *Preserving Nature.*
24. Clarke, "Grizzlies and Tourists", 26.
25. Clarke, "Grizzlies and Tourists", 28, and Jabin, "Grizzlies Turn Killer".
26. Clarke, "Grizzlies and Tourists", 29
27. Wikipedia contributors, "List of Fatal Bear Attacks".
28. The United States Department of the Interior, "Bear Management Glacier National Park", 1.
29. Department of the Interior, "Bear Management Glacier National Park", 15.
30. Kendall et al., "Grizzly Bear Density".
31. Leopold, "Widlife Management Report", 5, 3.
32. Kendall et al., "Grizzly Bear Density", 1694.
33. Donaldson and Kymlicka, *Zoopolis*, 3.
34. Donaldson and Kymlicka, *Zoopolis*, 3–4.
35. Nussbaum, "The Moral Status of Animals", 35.
36. Nussbaum, "The Moral Status of Animals", 32.
37. Olsen, *The Night of the Grizzlies.*

38. Olsen, *The Night of the Grizzlies*, loc 1595.
39. Olsen, *The Night of the Grizzlies*, loc 1817.
40. See e.g. "Grizzlies Turn Killer" and "List of Fatal Bear Attacks".
41. Bieder, *Bear*.
42. Haraway, *When Species Meet*, 16
43. Haraway, *When Species Meet*, 164
44. Haraway, *When Species Meet*, 16
45. Haraway, *When Species Meet*, 4
46. Westlund, "Training Is Enrichment".
47. Kolber, "Standing Upright".
48. French, "Experts Agree Bear's Behavior Was Unusual". After a 2010 mauling in Montana, the guilty sow was killed, but her cubs who had been on site, were moved to a zoo.
49. "Grizzly Mauls Photographer to Death". In a Glacier bear incident in 1987 that was a possible inspiration for the one in the novel, Charles Gibbs was killed by a grizzly sow he was photographing. The sow was not destroyed since she was deemed to have acted in defence of her cubs,
50. Westlund, "Training Is Enrichment".

Works Cited

Auden, Wystan Hugh. "The Guilty Vicarage: Notes on the Detective Story, by an Addict." *Harper's Magazine*, 1948, 406–12. https://harpers.org/archive/1948/05/the-guilty-vicarage/. Accessed 25 January 2020.

Barr, Nevada. *Track of the Cat*. New York: Berkley, 2003. 1993.

Barr, Nevada. *Blood Lure*. New York: Berkley, 2002.

Barr, Nevada. *Winter Study*. New York: Berkley, 2008.

Bertens, Hans, and Theo D'Haen. *Contemporary American Crime Fiction*. Basingstoke: Palgrave, 2001.

Bieder, Robert E. *Bear*. Reaktion Books, 2005.

Clarke, Jeanne. "Grizzlies and Tourists." *Society* 27, no. 2 (1990): 23–32. doi:https://doi.org/10.1007/BF02695481, https://link.springer.com/content/pdf/10.1007%2FBF02695481.pdf.

Donaldson, Sue, and Will Kymlicka. *Zoopolis: A Political Theory of Animal Rights*. Oxford: Oxford University Press, 2011.

French, Brett. "Experts Agree Bear's Behavior Was Unusual." *Billing's Gazette e-edition*, 29 July 2010. billingsgazette.com/news/state-and-regional/montana/experts-agree-bear-s-behavior-was-unusual/article_5cb13602-9b79-11df-b287-001cc4c03286.html. Accessed 30 March 2018.

"Grizzly Mauls Photographer to Death." *United Press International Archives*. billingsgazette.com/news/state-and-regional/montana/experts-agree-bear-s-behavior-was-unusual/article_5cb13602-9b79-11df-b287-001cc4c03286.html. Accessed 25 March 2018.

Haraway, Donna Jeanne. *When Species Meet*. Minneapolis: University of Minnesota Press, 2008.

Horsley, Lee. *Twentieth Century Crime Fiction* (Oxford: Oxford University Press, 2005).

Jabin, Clyde. "Grizzlies Turn Killer... Two Girls Die, Youth Mauled." *The Windsor Star* 1967.news.google.com/newspapers?id=XzY_AAAAIBAJ&sjid=m1EMAAAAIBAJ&dq=michelle%20koons&pg=4138%2C6134028. Accessed 25 March 2018.

Kendall, Katherine C., Jeffrey B. Stetz, David A. Roon, Lisette P. Waits, John B. Boulanger, and David Paetkau. "Grizzly Bear Density in Glacier National Park, Montana." *Journal of Wildlife Management* 72, no. 8 (2008): 1693–705. doi:https://doi.org/10.2193/2008-007.

Knight, Stephen. *Form and Ideology in Crime Fiction*. Bloomington: Indiana University Press, 1980.

Kolber, Adam. "Standing Upright: The Moral and Legal Standing of Humans and Other Apes." *Stanford Law Review* 54, no. 1 (2001): 163–204. doi:https://doi.org/10.2307/1229426.

Leopold, Aldo Starker. "Wildlife Management in the National Parks." US National Park Service, 1963. Reprint, 1969.

Messent, Peter. *The Crime Fiction Handbook*. Hoboken: Wiley, 2012.

Molander Danielsson, Karin. *The Dynamic Detective: Special Interest and Seriality in Contemporary Detective Series*. Uppsala: Acta Universitatis Upsaliensis, 2002.

Nussbaum, Martha C. *Love's Knowledge: Essays on Philososphy and Literature*. New York: Oxford University Press, 1990.

Nussbaum, Martha C. "The Moral Status of Animals." In *The Animals Reader: The Essential Classic and Contemporary Writings*, edited by Linda Kalof, Fitzgerald, Amy J. Oxford: Berg, 2007, 30–6.

Olsen, Jack. *The Night of the Grizzlies*. Crime Rant Classics 2014. Kindle, New York: Putnam, 1969.

Porter, Dennis. *The Pursuit of Crime*. New Haven: Yale University Press, 1981.

Sellars, Richard West. *Preserving Nature in the National Parks: A History - with a New Preface and Epilogue*. New Haven: Yale University Press, 1997.

The United States Department of the Interior. "Bear Management Guidelines, Glacier National Park." National Park Service, 2001.

Walton, Priscilla L., and Manina Jones. *Detective Agency: Women Rewriting the Hard-Boiled Tradition*. Berkeley: University of California Press, 1999.

Westlund, Karolina. "Training Is Enrichment-and Beyond." *Applied Animal Behaviour Science* 152 (2014): 1–6.

Wikipedia contributors. "List of Fatal Bear Attacks in North America" Wikipedia, the Free Encyclopedia. https://en.wikipedia.org/w/index.php?title=List_of_fatal_bear_attacks_in_North_America&oldid=937106579

Laboratory Tech-Noir: Genre, Narrative Form, and the Literary Model Organism in Jay Hosking's *Three Years with the Rat*

Jordan Sheridan

Jay Hosking's *Three Years with the Rat* tells the story of a listless young man who, after dropping out of university for the third time, returns home to Toronto in order to reconnect with his sister Grace and her partner John. He begins to settle into his new life, but soon Grace becomes increasingly anxious, agitated, and disappears without a trace. Shortly after, John also disappears. As the plot unfolds, we learn that Grace and John are experimental physicists who have been secretly developing a black box that allows them to travel to alternate dimensions. The novel follows the unnamed narrator as he performs the role of a private investigator by tracking down leads and discovering clues about Grace and John's disappearances, their secret experiments, and their time-bending black box. Central to the search is Buddy the laboratory rat, the last survivor of a group of rats with telemetry devices implanted in their bellies that track their location in alternate dimensions. Eventually, the protagonist's

J. Sheridan (✉)
New York University, New York, NY, USA

R. Hawthorn, J. Miller (eds.), *Animals in Detective Fiction*, Palgrave Studies in Animals and Literature,
https://doi.org/10.1007/978-3-031-09241-1_7

search yields a significant lead when Buddy disappears inside the mysterious black box only to return a week later with a note tied around his neck that reads, "This is the only way back".[1]

As the only form of communication between the narrator and his sister, Buddy functions as an epistemic shuttle between the narrator, his search for answers, and his lost sister in a way that mirrors how animals are used in laboratories to model the human body in experimental science.[2] However, *Three Years with the Rat* resists depicting Buddy and the narrator strictly in terms of what Donna Haraway calls "instrumental relations" between model organisms and the scientist by representing Buddy as a complex, sentient being who is worthy of care, consideration, and mourning.[3] Hosking's experience as a neuroscientist provides insight into the novel's depiction of Buddy, for (as his publication titles attest) his research has primarily involved the use of rats as model organisms in behavioural experiments that investigate decision-making processes and behaviours associated with addiction and gambling.[4] In fact, in a segment of "Clues to My Crime", a literary review blog where authors explain the influences behind their work, Hosking poses a series of questions that challenge preconceived notions of laboratory rats:

> Rats are far smarter than you might think, and working with them has been the highlight of my scientific career. Did you know that you can teach rats to gamble (and that they're about as good/bad at it as we are)? Did you know that they enjoy being tickled? Did you know that they display empathy, and will even give up chocolate to help another rat in distress? Rats share a lot of qualities with my favourite people: they're curious, resilient, motivated, unglamorous, clever, they bond well and, more than anything else, they survive against all odds.[5]

Hosking employs something approaching what Frans de Waal calls "heuristic anthropomorphism" to challenge preconceived notions of rats by comparing their abilities and traits to humans. de Waal argues that this type of anthropomorphism is useful for scientists because it offers a way to "reorient" scientists' "views and provide new frameworks for research" by allowing scientists to explore ideas normally considered to be outside scientific epistemology.[6] While it must be noted that this comparison is framed by Hosking's knowledge and experience of rats as experimental objects, his comparison encapsulates the central question at the heart of this chapter: by offering alternative ways of thinking about model organisms outside the strict confines of the laboratory, how does literature challenge narratives of model organisms as inert, passive scientific objects? I suggest that *Three Years with the Rat* challenges the epistemic and temporal structures that model organisms provide for the biological sciences as "spaces of representation" for human illness by mapping the narrative framework of model organism experiments onto a hybrid science fiction detective plot, or tech-noir, that centres around a laboratory rat. [7] The question of genre offers a particularly effective way to understand how *Three Years with the Rat* utilizes model organisms at both the level of plot and story because Hosking's complicated narrative structure highlights the work that model organisms do within experimental science to produce and circulate narratives of the human body. [8]

Emily Auger defines the genre of tech-noir as a text that depicts technology "as a destructive and dystopian force that threatens every aspect of our reality" and "expose[s] the temporal nature of concepts of identity and society" by showing us that our contemporary anthropocentric and human exceptionalist social orders are not impervious.[9] Auger's definition is important for a discussion of *Three Years with the Rat* because she emphasizes a clear break between detective fiction and tech-noir. "Unlike detective fiction", Auger argues, tech-noir addresses "all-encompassing rather than isolated problems such that the entire worldview and physical environment [...] are compromised by or suffer from the aftermath of a violent revision of what we currently take to be reality".[10] In other words, what separates detective fiction from tech-noir is the scope of each genre's philosophical and sociological critique of the relationship between science, history, and the individual. Auger argues that in tech-noir narratives, the story takes place in a science fictional world defined by dystopian developments in science and technology; in detective fiction, the world is defined by an individual's perception of the state of her/his own personal existence in relation to the contemporary world.[11] However, if we consider classic tech-noir texts like *The Terminator* (1984), *Do Androids Dream of Electric Sheep* (1968), or *Logan's Run* (1967), Auger's distinction between the "all-encompassing" and the "isolated" does not hold up because each in its own way tells the story of an individual who struggles with the threat of technology and scientific advancement on a personal level. Instead of a strict demarcation between tech-noir and detective fiction we can say that there are two levels of symbolic alienation of the human subject by the technologization of society occurring in tech-noir narratives—or rather that tech-noir situates the detective narrative within a science fiction world that has been disrupted by technological advancement.

Franco Moretti argues that science has always been fundamental to detective fiction: Science is not simply "part of" detective fiction, but rather that it "enters" the genre as a "semiotic mechanism (the decipherment of clues)" that attaches ideological formations to certain people, objects, and spaces.[12] "The decipherment of clues", for Moretti, "presupposes that 'science' is identified with an organicist ideology based on the 'common-sense' notion that differences in [social] status cannot be altered".[13] Thus in Moretti's view, the resolution of the mystery often found in detective narratives positions science in a "stabilizing role", or at least keeps the relationship between "science and society *unproblematic*".[14] Within this reading of the detective genre, Moretti claims that the role of science and thus detective fiction is essentially conservative because it "creates a problem, a 'concrete effect'—the crime—and declares a sole cause relevant: the criminal. It slights other causes (why is the criminal such?) and dispels the doubt that every choice is partial and subjective".[15] Armed with the science of detection, the detective then functions as an alibi for the exploitative, hierarchical social order of contemporary society by singling out the criminal. Thus, Moretti argues that, "in finding one solution that is valid for all—detective fiction does not permit alternative readings—society posits its unity, and, again, declares itself innocent".[16] Moretti's critique of the detective genre is useful because he offers a way to think about how science functions ideologically in genres that are more scientific such as tech-noir. Tech-noir features a more explicit and ambivalent treatment of science, representing it as both disruptive and stabilizing at the same time. By configuring Buddy as an individual and a scientific tool for the decipherment of clues, *Three Years with the Rat* exemplifies this kind of negotiation of disruption and stabilization because as a model organism his body has been altered to represent the human body, but the biotechnologies that came together to make this possible are central to the kinds of alienation that tech-noir literature depicts.

Hosking's novel appears at a time when the field of science and technology studies has begun to question the roles that non-human animals play in the biological sciences as model organisms for human illness and disease. In its broadest definition, the term "model organism" refers to "nonhuman species that are extensively studied in order to understand a range of biological phenomena, with the hope that data, models, and theories generated will be applicable to other organisms".[17] This definition is useful because it highlights the temporal and speculative structure of model organism research that applies knowledge about animal bodies onto future human bodies. Model organisms can thus be thought of as "heuristic devices" because although their bodies are *similar* to human bodies, they can never be *identical*.[18] The heurism of the model organism is reminiscent of de Waal's notion of "heuristic anthropomorphism" noted above in that the model organism represents a strategic mapping of similar but not identical physiologies of humans and other animals onto one another.[19] It allows scientists to test hypotheses about the human body without the presence of a human body. However, as Nicole Nelson argues, the heuristic nature of the model organism depends on "epistemic scaffolds" that involve a "process of building up a structure of evidence and arguments to make claims" about any given experimental procedure (such as using a maze to understand stress and anxiety in rodents).[20] Epistemic scaffolds are built from a variety of evidence and arguments—from studies positing the efficacy of experimental tools and apparatuses, to studies on the biological links between certain species of animals such as humans and rats.

One of the major aims of postgenomic sciences is to make animal bodies as close as possible to human bodies through specialized breeding practices and genetic editing—a process Gail Davies calls "corporeal equivalence".[21] Perhaps more importantly for a discussion of model organisms in literature, Davies argues that corporeal equivalence has a narrative component as well, both in and outside the laboratory. Narratives of equivalence bring humans and animals together by organizing certain types of genetic similarities according to their experimental applications.[22] Furthermore, as Davies points out, relationships between model organisms and hypothetical human contexts "are not fixed, but performed, in part through the narratives that are told about them. Narratives in science, as elsewhere, are important as they order histories and point towards futures: framing temporality, allocating cause and effect".[23] These animals "frame temporality" and "allocate cause and effect" through the logic of the experiment that maps animal bodies (in the present) onto human bodies (in the future).[24] Then, after this mapping is complete, animal bodies disappear because they are designed to refer to the human body rather than their own.

It is in this narrative which Hosking's novel intervenes; throughout the novel, Buddy frames the temporality of the story by allocating the cause and effect through the narrator's use of scientific practices and epistemology. The narrator learns that he has both to reject Grace and John's instrumentalist understandings of model organisms and pay attention to Buddy as an individual and, in fact, allow Buddy to guide him through the alternating time-scape. However, at the same time, the novel also positions Buddy's body as a crucial clue that leads the narrator to Grace and John because the narrator must engage with Grace's scientific methodologies to use Buddy as a model organism. In this chapter, I show how the novel holds up these two seemingly conflicting understandings of Buddy to present a more complicated narrative of the relationship between the model organism and the scientist. I argue that the novel builds this critique first through a series of critical encounters with Grace and John where they try to teach the narrator how to use model organisms as scientific tools, and second, by depicting the narrator's scientific detection as a less anthropocentric and more empathic way of conducting scientific research.

Grace, 2006: The Model Organism, Scientific Epistemology, and Mystery

One month before Grace disappears, she gives the narrator a tour of "*The Center for Animal Modeling*" where she works (128). These scenes serve as the foundation for the text's critique of the model organism, both outlining how Grace discovers the alternative dimension by using model organisms and establishing the narrator's point of view as a critical perspective. As the narrator moves through and interacts with the laboratory space, he exposes scientific discourses of animal life that make them into objects by closing off their individuality and limiting human interaction.

From the moment that Grace and the narrator reach the laboratory, the narrator begins to question the basis of scientific epistemologies that rely on model organisms. "This is your top secret facility?" the narrator mockingly asks Grace, "Animal modelling? Beauty queen rats, in a scummy ally around the corner from the convenience store" (128)? By humorously contrasting the "animal model" as it appears in the name of Grace's laboratory with the notion of the fashion model, the narrator questions the structure of the scientific model. Recalling the brief outline of the model organism I offered above, model organisms "model" human bodies through a complex association of the human body and the bodies of other animals based on genetic commonalities. The narrator's comparison between fashion models and model organisms calls attention to the model organism as an embodied form of representation. The joke relies on the reader's understanding that in the case of the fashion model, the model's body serves as a platform for representing how a piece or ensemble of clothing fits or wears on an "ideal" human body. The joke evokes questions about what exactly a model organism "models" and challenges us to consider how model organisms' bodies become a form of representation. As Ian Hacking suggests, scientific models "are doubly models. They are models of the phenomena, and they are models of the theory".[25] Hacking argues that we can think about models as "intermediaries" between scientific theories and the world because they simplify and isolate phenomena, allowing them to be more easily represented.[26] Similarly, Hans-Jörge Rheinberger argues that, "the process of modelling is one of shuttling back and forth between different spaces of representation".[27] In *Three Years with the Rat*, the narrator's joke draws attention to the representational form of the model organism by undermining scientific narratives of corporeal equivalence that obscure the central role of representation in experimental practices. By humorously evoking model organism representation, the text challenges scientific discourses that reduce animals to conduits of biological and physiological processes that collapse human and animal bodies.

After winding through a labyrinth of window-lined hallways, they reach a scrub room where they wash their hands and put on protective overalls before entering the animal house that is kept in darkness to maintain rats' nocturnal circadian rhythm:

> Without vision, my ears became extremely sensitive, and I could hear the soft rustling sound of small animals. Grace was silent. I imagined her standing still a few steps in front of me, smug, letting this moment of pure darkness linger. Soon the animal rustling became scratching and then clanging on metal. The experience was disorienting, and I reached out for the cool comfort of a wall. I still couldn't hear Grace or even sense her in the room (130–131).

This scene metonymically links the narrator, murine sensory experience, and mystery through a symbolic "pure darkness" oriented towards Grace's spectral presence (130). His confusion and disorientation as he moves through the room foreshadows his intellectual experience of being in the dark tracking down Grace and John after they disappear. This association also links the narrator to the rats, challenging us to think about how rats experience the world and positions the narrator as someone who sees rats as complex sentient beings. The text solidifies the critical potential of this perspective when Grace turns the lights back on and the narrator begins to make connections between his sight and other senses:

> Though it was still hard to see, I could make out the stacks of cages that filled the room. Inside each cage were two pairs of black beady eyes that seemed to stare back at me. The sounds of the room made sense now: the rats were pushing against the lids with their forepaws, or chewing on the edge of plastic tubes, or digging through the soft bedding that lined the floor of their homes. The whole room felt boxed, but alive (131).

With the lights on, the narrator has a different experience; now he sees the room in all its living complexity. In a similar way that the "darkness" metonymically associates the narrator with the rats, "lightness" associates his perspective with an awareness of how the physical and social structures that contain and control rats within a laboratory come together to make them into model organisms. In meeting the rats' gaze, the narrator becomes aware of the liveliness of the rats as such. This scene foreshadows the narrator's eventual realization that he will have to work with Buddy and trust him less like a scientific tool and more like a complex individual. This reflects Haraway's argument that animals used in research are "material-semiotic actors" rather than "pre-discursive bodies just waiting to validate or invalidate some discursive practice".[28] Haraway refers to animal behavioural science and primatology where scientists read and interpret complex material social interactions and then theorize their meanings. Haraway argues that this process requires active participation on the behalf of the observed animals rather than a scientist's mind alone. Moreover, the final contrast "boxed, but alive" highlights the extent to which the geography of the laboratory is comprised of a series of spaces that are designed to facilitate the "transformation" of the animal body into a scientific object.[29] The rats are indeed alive, and it is precisely their liveliness that needs to be boxed in, modelled, and capitalized on.

The text contrasts the narrator's description of the rats in the next scene when they move into the final space of the laboratory containing Grace's mysterious black box and specialized rats with telemetry devices implanted in their bellies (132). She begins to explain that she uses classic Pavlovian conditioning to make the rats interact with the black box and signify their experience of time by raising a paw. However, before Grace can finish, the narrator intervenes by asking, "how do you really know what is going on for the rat" (133)? She explains that she conditioned the rats in a way that would allow her to determine whether or not their expectations of a pre-determined set of time matched an objective measurement of time. The narrator rejects Grace's explanation, arguing, "no matter how you manipulate these rats, you can't take out the part where they all live in the real, objective world" (135). This disagreement about "what is going on for the rat" echoes Vinciane Despret's argument that scientific epistemology actively resists asking questions about model organism agency (133). Instead, she argues, experiments are designed to exclude such questions by ruling out any response not predetermined by the scientist. Scientific epistemology excludes animals' interests on the basis that such consideration would be anthropomorphic. However, as Despret argues, behavioural science doesn't necessarily "restrict" anthropomorphism per se, but rather makes it imperceptible and "invisible" by framing experiments in such a way that it "blocks the possibility that the animal could show how he takes a position with respect to what is asked of him".[30] The narrator's question sets the stage for the relationship he develops with Buddy based on a non-anthropocentric form of relationality where the narrator requires Buddy's perspective in order to find his way through the alternate dimensions.

John, 2007: Corporeal Equivalence and Empathy

Shortly after Grace disappears, the narrator turns to her partner John for support but instead finds him elusive and absorbed by a mysterious project. After a few weeks, the narrator pushes John to open up about his project and demands to know why he refuses to speak of Grace's disappearance. In response, John invites the narrator to his apartment where he reveals that he stole three of Grace's rats from the *Center for Animal Modeling* and replicated her black box. This section of the novel stages a series of transitional moments for the narrator; it builds off many of the ideas about scientific epistemology and model organisms established through Grace's tour of the laboratory by putting them into dialogue with more normative discourses of domesticated animals and the space of the home. Through the tour, the text exposes the narrative elements of corporeal equivalence first by showing how laboratory rats are a domesticated, companion species; second, the text establishes a model relationship between the human characters and the rats by staging a scene where John and the narrator name the rats after themselves and Grace before sending them into the black box.

After an initial moment of hesitancy with the stolen rats, the narrator describes how he began to recognize similarities between rats and other domesticated animals: "In a way, they reminded me of dogs", he admits, "I even let one crawl up onto my shoulder, its little nails like little pin-pricks across my skin, its whiskers and nose tickling my neck. I dreamt of that sensation for days. One evening, while relaxing over scotch and pop-corn, I gave one of the rats a kernel. It held it in its forepaws and ate it thoughtfully and I couldn't help thinking it was happy" (117). This scene is in stark contrast to the narrator's experience at the *Center for Animal Modeling* that introduced him to rats solely within institutional structures of the laboratory. Here, the narrator interacts with them in a more personal manner where the rat is outside its cage and walks on his body. This bodily connection is key to this scene because it subverts the logic of corporeal equivalence that creates bodily relationality in order to instrumentalize the body of the animal. Instead, the narrator's bodily connection with the rat causes him to empathize with it and to consider its welfare as an individual, building on his challenge of Grace's scientific epistemology in the previous scene.

The novel subverts this logic further in the next scene where John and the narrator name the stolen rats: "the largest of them was *Little John* and the fat one with the fewest black markings was *Little Grace*. Although all three were males, we still referred to Little Grace as *her* and *she*. John offered to name the third rat, the runt, after me. 'Not a fucking chance', I told him. 'Call him Buddy'" (118). By naming the rats after the human characters, the text performs a kind of mock narrative of equivalence that positions the rats as models for the human characters once, without the narrator's knowledge, John begins to send the rats into the black box causing their health to decline. Together, the rats embody a *mise en abyme* by reflecting the narrative structure of the novel. This *mise en abyme* constitutes a "story within a story" that resembles an experiment: John uses the rats as models by sending them into the box and recording the data he reads from their telemetry devices. However, the novel resists any straightforward reading of the rats as model organisms through the narrator's empathetic response to Little Grace, the "first [rat] to show signs of distress" (118). The narrator recalls how one day upon visiting John he heard a concerning sound emanating from John's apartment, then when John opened the door "the source became clear: one of the rats was shrieking in their cage. Little Grace was in the corner and raised up onto her back two legs. Her front legs were outstretched like arms and her teeth were bared as if to say, *Stay away from me*" (118–119). Though the rat's body has no material connection to Grace's body, the name "Little Grace" links the rat to the narrator's sister and calls on the reader to associate Little Grace's emotional state with Grace's emotional state before she disappeared. This connection challenges the reader to consider the rat's pain as analogous to human pain by using common detective fiction genre conventions that position women as victims of violence that the male detective figure must save, assist, or seek justice on behalf of.

Two days later, when John notifies the narrator of Little Grace's death, the narrator rushes over to his apartment to find her body "laid out" on a newspaper, "camouflaged" against the text (120). "There seemed to be nothing wrong with her", the narrator describes, "only that life had somehow drained out of her and left her rigid and frozen in a snarl. John was inconsolable, his gaze blank, absent. [...] Eventually I convinced him to leave the apartment for a walk. We brought an arrowhead shovel and buried Little Grace in the park (129). While John's mourning follows the *mise en abyme* structure of the rats' journey into the box, it also opens up questions of mourning and grieveability. Is Little Grace worthy of mourning? Does she "matter" beyond her symbolic equivalence with Grace? By describing Little Grace as "camouflaged", the text draws explicit attention to the narrative and textual basis of the model organism whose meaning and value within experiments is evaluated by the narrative of equivalence generated by the experiment. By walking the reader through a view of the model organism as a domesticated animal and performing a literary version of a narrative of equivalence, the novel suggests that the corporeal equivalence of experiments can be a source of empathy and connectivity between humans and animals.

Buddy, 2008: Beyond Laboratory Labour as Model Organism Agency

The "September: The City, 2008" section of the novel takes place after both Grace and John have disappeared and the narrator fully takes on the role of private detective and inherits Buddy as a companion, after he discovers him alone inside John's black box. This section builds on the novel's critique of model organism epistemology by bringing together the two conceptions of laboratory rats developed in the previous sections into dialogue. The narrator is forced to challenge his empathetic view of Buddy as a companion and to engage with him in a way that approaches Grace and John's instrumentalism. Instead, the narrator develops a hybrid form of relating to Buddy that acknowledges his sentience as well as the fact that his body is a source of important information.

One night shortly after being reunited with Buddy, the narrator loses him inside the black box while preparing food:

> He stands on my forearm without complaint and watches as I pick up my scotch. [...] I remember John feeding hard little cylinders to the rats but I have no idea what the pellets actually were. I consider my current options, pasta or toast, and put a slice of bread in the toaster oven [...] While the bread is toasting, I put Buddy on the top of the miniature box. [...] and then I hear a scrabbling sound and look to the box in time to see Buddy's tail disappear (44).

This scene bears a striking similarity to the passage that describes the narrator's change of heart towards the rats in the 2007 timeline; as the previous scene describes the narrator and Buddy bonding over shared food, here Buddy enters the box while the narrator tries to feed him. Buddy's escape within this domestic setting links the ideas of care, consideration, and empathy developed in the previous section to scientific epistemology. In other words, it is Buddy's companion-ness that allows him to escape and thus makes him a good model organism: Companion-ness and model-ness align in this scene not in spite of positive animal ethics but because of them.

One night a week later, the narrator hears a crash in the kitchen and finds Buddy amongst the wreckage of John's black box. Then, when he "reaches down" to put Buddy back into his cage, he feels a small "lump" in Buddy's side (57). Instantly, the narrator realizes that this lump is a telemetry device, "implanted two years ago by Grace and John" (57). In this scene, the text aligns the narrator's tenderness and his concern with the recognition that Buddy is a bearer of crucial information. Similar to the previous passage where he "uses his finger" to pet Buddy, here he uses his finger to discover that he has been a model organism this whole time. This discovery is his first clue: that Buddy's body is the missing piece of the puzzle that he has been searching for, giving him a new direction and a renewed sense of purpose.

The narrator's discovery leads him back to *The Center for Animal Modeling* where he tries to extract Buddy's data with Grace's equipment. The narrator's subversive use of Grace's equipment reflects his distance from scientific epistemology and his attitude towards Buddy. Echoing on the scene discussed above where the narrator realizes that the rats in the laboratory are "boxed, but alive" (131) the narrator understands Buddy in a way that approaches what Donna Haraway calls "unfree partners" who are "crucial to the work of the lab and, indeed, are partly constructed by the work of the lab".[31] The notion of unfree partners builds off Haraway's investigation of animal behavioural science and argues for the need to understand laboratory animals in terms of labour rather than absolute victimhood. In a similar way that animals in behaviour science are "material semiotic actors" and participants, laboratory animals "are not 'unfree' in some abstract and transcendental sense. Indeed, they have many degrees of freedom in a more mundane sense, including the inability of experiments to work if animals and other organisms do not cooperate".[32] In other words, the material limitations placed in laboratory animals in cages and laboratory spaces does not necessarily mean that animals are devoid of agency. Haraway's argument is a compelling place to begin to think outside of the logic of absolute subjugation that offers little room to imagine animals like Buddy in terms other than death and suffering. Laboratory animals participate in the production of scientific knowledge through their cooperation, and for this reason they have a peculiar form of limited agency.

As Zipporah Weisberg points out, Haraway's theories of companionship and partnership do little to wrest laboratory animals from "the humanist project of domination she ostensibly disavows".[33] However, this critique doesn't mean that considerations of how animals live within the decidedly and unapologetically humanist spaces like laboratories should be rejected because they don't adopt liberationist perspectives that understand animal freedom in absolute terms alone. Similarly, Hosking's novel functions as an exploration of animal agency and sentience from *within* the framework of the laboratory and scientific epistemology by offering ways to unsettle discourses of mastery and total control within experimental science.

In *Three Years With The Rat*, once the narrator reaches the procedure room, he boots up the computer and writes a note for Grace and John: "*I'm coming for you. Hold on*" (95). As he attaches the note to Buddy's body, the narrator whispers, "Bring me back something useful, Buddy" (95). In this scene, Buddy is a companion, a helper, and a clue combined. Moments after Buddy disappears into the box a graduate student approaches: "Did you put an animal in there?" She asks, "'*What* animal did you put in there' [...] 'We always just called him Buddy'" the narrator replies (98). The following exchange complicates Haraway's theory of "unfree partners" by showing how model organism agency greatly depends on how the humans around them read and interpret their participation in experiments. Soon after the narrator's response, the graduate student asks, "What could you possibly need him for? His data?" and then offers to "copy of all the data" in exchange for Buddy (98). This proposition frames Buddy as something that has value for its ability to house or produce information. The graduate student assumes that the narrator could not *need* Buddy because he does not *need* the information contained within his body.

While the narrator certainly does *need* Buddy's information, his understanding of that need is much more complicated, for as he reflects, "her offer is exactly what I came for but something doesn't feel right. Deep in me is a gnawing sensation. It takes a moment for it to come to the surface", and he asks, "'What will you do with him?' 'The animal? Look at its data, sacrifice it, see what's possibly different about this particular subject'" (98). What "doesn't feel right" for the narrator is that his feelings for Buddy conflict with his desire to find his sister because as I have been outlining—the narrator closely associates Buddy with Grace and John. "Sacrifice it. Kill Buddy [...] I can't let you do that", the narrator responds (98). By subverting the graduate student's euphemistic reference to Buddy's death, he hits a nerve and she threatens him by replicating a string of scientific justifications for her use of model organisms such as the fact that he was "bred for one reason", that he "belongs to this lab", and is thus "stolen property" (99). The narrator's answer is simply to state: "He's my last connection" (99). This configuration of Buddy as a "connection" amends Haraway's theory of agency by showing that the problem with conceptualizing model organism agency as a product of scientific labour is that it orients animal life around what they can *do* for science. By rejecting the graduate student's reference to Buddy's death in relation to his use-value, the narrator ascribes value to Buddy outside of the strict confines of epistemological production. However, questions remain: why "use" Buddy at all? Why doesn't the narrator take the risk and enter the black box without the reassurance that Buddy's safe passage offers? Is the connection Buddy offers a way for the narrator to mourn for his missing sister, and does this then form a complicated link between grief, the human, and the model organism?

The Other Side: Sacrificial Companionship and Mourning

Though Hosking leaves such questions largely unanswered, I suggest in conclusion that the final scenes of the novel advance a productively nuanced view of model organism science and thus the relationship between the model organism and the scientist. On the one hand, the novel goes to great lengths to demonstrate the importance of understanding model organisms as sentient beings whose lives matter. On the other hand, sometimes research is necessary to save human lives as well. By the end of the novel, it becomes clear that Hosking's critique of model organism science has more to do with challenging perceptions of model organisms than upending the institutional apparatus of the model organism industry. Instead, he focuses on the fractured relationship between the scientist and the animal that can be based on care, consideration, and welfare even though it inevitably contains a great deal of violence towards the animal.

In the final section of *Three Years with the Rat*, Buddy and the narrator cross to another dimension that the text refers to as "the other side" (231). This section starts where the previous section closed. After momentarily losing sight of Buddy, the narrator finds him waiting near a clearing in the trees, as the narrator recalls, "I bend forward to pick him up and he runs out of reach. I take a step forward and he runs ahead again. We repeat this process a few times more before I recognize the pattern. I am being led" (234). This reflection outlines the premise to the concluding section of the novel where Buddy leads the narrator to John and Grace by training the narrator to follow him. As can be expected from the title, "The Other Side", here Buddy's training of the narrator is a mirror opposite of the kind of laboratory conditioning that Grace used to train her rats to measure subjective time. The narrator's following of Buddy in this section is the culmination of the subversive logic of the narrator's approach to model organisms as being more than simple tools for research.

However, Buddy's time as a guide to the narrator ends when they are separated as they flee the mysterious figure called "the hunter" that has been following them (235). After being finally reunited with John, he explains that Grace is nearby but cautions him that she is unwell. When the narrator finds Grace, he discovers that she is not alone but surrounded by "a sea of people" that are all versions of "Grace" from different dimensions, come together to solve the problem of subjective time (252). Grace stubbornly refuses to leave, and the narrator and John decide to leave without her. Before long they hear the sound of "a scared rat" and realize that the hunter has Buddy (250). When the narrator confronts the mysterious figure, he discovers that the hunter is a version of himself from another dimension and after pleading with himself to spare Buddy, his alternate self slams Buddy against the boat and watches as "the poor rat bounces off the hull, goes limp" (262).

While Buddy survives this attack, it is his last scene; the narrator and John escape the hunter and after they arrive back to "the other side" they find out that "Grace's body was discovered over Labour Day weekend September 2008", four months before the narrator reached the other side with Buddy (265). Buddy dies shortly after the narrator attends Grace's funeral, and he buries him with his fellow rats, Little John and Little Grace. What conclusion can we draw from Buddy's sudden and rather anti-climactic end? Within Moretti's critique of the detective novel, Buddy's role as a guide who helps to decipher clues keeps "science and society *unproblematic*" by making sure that normative society remains intact.[34] In *Three Years With The Rat,* this formula is slightly askew because Grace's science itself is the major disrupting force in the novel, however, given that a large part of how the narrator solves the mystery of her disappearance both relies on that same science, it is clear not all science is created equal. By positioning Grace and the narrator at opposite ends of an ethical spectrum of scientific practice, the text puts forward the latter as something that has the ability to stabilize the other. But where exactly does this leave the rats?

The text leaves this question mostly unresolved because after all Little John, Little Grace, and Buddy all die by the end of the novel. This ambivalence stems from the fact that the concept of the model organism structures much of the plot. Because of this, *Three Years With The Rat* ultimately mirrors narratives of experimental sciences that position the deaths of model organisms as pedagogical events for the human characters—even if these events were meaningful and the animal models were cared for and mourned. The model organism is a companion, but one that remains sacrificial. However, thinking back to Buddy's final scene, that the attack occurs at the hands of the narrator's alternate self suggests that while scientists cannot situate themselves outside of the violence of experimentation, they can make life more bearable for animals by taking a more empathetic approach to their well-being. The narrator chooses to fight and kill his alternate self to save Buddy from dying in the other side. This choice suggests that it is possible for model organisms to remain individuals and thus not be completely subsumed by scientific epistemology. Read this way, the narrator's last mention of Buddy is revealing: "Buddy had been through so much that it seemed a waste to die now. And while I hadn't been able to cry at my sister's funeral, for some reason I found myself wrenching out hot, angry tears for my lost companion" (267). As one of the last scenes in the novel, the narrator's mourning of Buddy pushes the reader to more directly engage with scientific discourses of model organisms as expendable and how they prevent our ability to make meaningful identifications with model organisms as complex beings who we are capable of connecting to on emotional and affective levels.

Notes

1. Hosking, *Three Years*, 12.
2. Rheinberger, *Epistemic Things*, 108.
3. Haraway, *Species Meet*, 71.
4. Hosking's scientific work primarily investigates the neurochemical basis of decision-making: "Sensitivity to Cognitive Effort Mediates Psychostimulant effects on a novel rodent cost/benefit decision-making task"; "Dopamine Antagonism Decreases Willingness to Expend Physical, But Not Cognitive, Effort: A Comparison of Two Rodent Cost/Benefit Decision-Making Tasks"; "Disadvantageous decision-making on a rodent gambling task is associated with increased motor impulsivity in a population of male rats".
5. Hosking, "Clues".
6. De Waal "Anthropomorphism and Anthropodenial", 268.
7. Rheinberger, *Epistemic Things*, 108.
8. *Three Years with the Rat* has five main sections that correspond to the months between August to December and spatial settings of The Apartment, The City, The Lab, The Suburbs, and The Box. Hosking divides each section into sub-sections that correspond to three descending years: 2008-2007-2006. Thus, the novel's structure relies heavily on contrasting narrative time (the time of the reader) and story time (the time of the characters). The narrative time of the novel moves forward through a sequence of reverse time-scales in a wave-like fashion that is mimetic of the temporality of the characters' experience of time-travel. As such, the text

organizes the story into four timelines, but since they are broken up in ascending order, the novel works backward and forward at the same time as a series of flashbacks until the final section that is "outside" of time, or rather, within a space of temporal convergence of all three timelines.

9. Auger, *Tech-Noir*, 21.
10. Auger, *Tech-Noir*, 21.
11. Auger, *Tech-Noir*, 21.
12. Moretti, *Signs*, 21.
13. Moretti, *Signs*, 21.
14. Moretti, *Signs*, 144.
15. Moretti, *Signs*, 144.
16. Moretti, *Signs*, 144.
17. Ankeny and Leonelli. "What's So Special", 320.
18. Shapiro. "A Rodent", 449.
19. De Waal "Anthropomorphism and Anthropodenial", 268.
20. Nelson Model Behavior, 86.
21. Davies "Humanized Mouse?", 4.
22. Davies. "Experimental Life", 131.
23. Davies. "Experimental Life", 130.
24. Ankeny et al. "Making Organisms", 486.
25. Hacking, *Representing*, 216.
26. Hacking, *Representing*, 217.
27. Rheinberger, *Epistemic Things*, 108.
28. Haraway, *Primate Visions*, 310.
29. Lynch, "Sacrifice", 267–70.
30. Despret, *Animals Say,* 91.
31. Haraway, *Species Meet*, 72.
32. Haraway, Species Meet, 72–73.
33. Weisberg, "The Broken Promises of Monsters", 23.
34. Moretti, *Signs*, 144.

Works Cited

Ankeny, Rachel A. and Sabina Leonelli. "What's so Special about Model Organisms," *Studies in the Philosophy of Science* 42, no 2 (2011): 313–323.

Ankeny, Rachel A. et al. "Making Organisms Model Human Behavior: Situated Models in North-American Alcohol Research, Since 1950," *Science in Context* 27, no. 3 (2014): 485–509.

Arluke, Arnold. "'We Build a Better Beagle': Fantastic Creatures in Lab Animal Ads," *Qualitative Sociology* 17, no. 2 (1994):143–158.

Auger, Emily. *Tech-Noir Film: A Theory of the Development of Popular Genres.* Intellect: Bristol, 2011.

Clause, Bonnie. "The Wistar Rat as a Right Choice: Establishing Mammalian Standards and the Ideal of a Standardized Mammal," *Journal of the History of Biology* 26, no 2 (1993): 329–349.

Davies, Gail. "Mobilizing Experimental Life: Spaces of Becoming with Mutant Mice," *Theory, Culture & Society* 30, no. 7-8 (2013):129–153.

Davies, Gail. "What is a Humanized Mouse? Remaking the Species and Spaces of Translational Medicine," *Body & Society 18*, no. 3–4 2012, pp. 126–155

de Roo, Brad. "Jay Hosking Interviewed." *Canadian Notes and Query.* 2016.

de Waal, Frans. "Anthropomorphism and Anthropodenial: Consistency in Our Thinking about Humans and Other Animals," *Philosophical Topics*, Vol. 27, No. 1 (1999): 255-280

Despret, Vinciane. *What Would Animals Say if We Asked the Right Questions?* Minneapolis: U of Minnesota P, 2012.

Hacking, Ian. *Representing and Intervening.* Cambridge: Cambridge UP, 1983.

Haraway, Donna. *Primate Visions: Gender, Race, and Nature in the World of Modern Science.* New York: Routledge, 1989.

Haraway, Donna. *When Species Meet.* Minneapolis: U of Minnesota P, 2008.

Hoffman, Josef. *Philosophies of Crime Fiction.* Trans. Carolyn Kelly, Nadia Majid, Johanna Da Rocha Abreu. Harpenden: No Exit, 2013.

Hosking, Jay. *Three Years with the Rat.* Toronto: Penguin Group, 2016.

Hosking, Jay. "Clues to My Crime," The Booklist Reader, accessed January 2017, https://www.booklistreader.com/2017/05/24/books-and-authors/clues-to-my-crime-jay-hoskings-three-years-with-the-rat/.

Logan, Cheryl. 'Are Norway Rats ... Things?': Diversity Versus Generality in the Use of Albino Rats in Experiments on Development and Sexuality". *Journal of the History of Biology* 34 (2001): 287–314.

Leonelli, Sabina and Ankeny, Rachel. "What Makes a Model Organism?" *Endeavor.* 39, no. 4 (2013): 209–212.

Lynch, Michael. "Sacrifice and Transformation of the Animal Body into a Scientific Object: Laboratory Culture and Ritual Practice in the Neurosciences," *Social Studies of Science* 18, no. 2 (1988): 265–89.

Moretti, Franco. *Signs Taken For Wonders: Essays in the Sociology of Literary Forms.* New York: Verso, 1997.

Nelson, Nicole. *Model Behavior: Animal Experiments, Complexity, and the genetics of Psychiatric Disorders.* U of Chicago P, 2018.

Rheinberger, Hans-Jörge. *Toward a History of Epistemic Things: Synthesizing Proteins in the Test Tube.* Stanford: Stanford UP, 1997.

Rzepka, Charles J. *Detective Fiction.* Cambridge: Polity, 2005.

Scaggs, John. *Crime Fiction.* New York: Routledge, 2005.

Shapiro, Kenneth. "A Rodent for Your Thoughts: The Social Construction of Animal Models," *Animals in Human Histories: The Mirror of Nature and Culture*, ed. by Marry J. Henniger-Vos. Rochester: U of Rochester P, 2002.

Sweeney, Susan E. "Crime in Postmodernist Fiction," *American Crime Fiction*. ed. Catherine Ross Nickerson. Cambridge: Cambridge UP, 2010.

Weisberg, Zipporah. "The Broken Promises of Monsters: Haraway, Animals and the Humanist Legacy," Journal for Critical Animal Studies, Vol. 7, No. 2 (2009): 22–62.

Reptiles, Buddhism, and Detection in John Burdett's *Bangkok 8*

Nicole Kenley

Introduction

John Burdett's Bangkok series, comprising six novels featuring Royal Thai Police Detective Sonchai Jitpleecheep, connects reptiles and Buddhism from its inception. The first line of the first novel, *Bangkok 8*, introduces these concepts together: "The African American marine in the gray Mercedes will soon die of bites from *Naja siamensis*, but we don't know that yet, Pichai and I (the future is impenetrable, says the Buddha)".[1] Here, the *Naja siamensis*, or Thai spitting cobra, is figured as a deadly, menacing force. By the novel's end, though, Jitpleecheep's understanding of cobras and other reptiles will transcend this initial portrayal, utilizing a deeper understanding of the ethics of karma to comprehend the value of reptiles as living beings beyond the threat posed by a deadly cobra. Burdett anthropomorphizes the reptiles in the novel, comparing them to human crime victims and thereby highlighting the value of reptile lives. This

N. Kenley (✉)
Baylor University, Waco, TX, USA
e-mail: Nicole_Kenley@baylor.edu

R. Hawthorn, J. Miller (eds.), *Animals in Detective Fiction*, Palgrave Studies in Animals and Literature,
https://doi.org/10.1007/978-3-031-09241-1_8

technique serves multiple purposes in the text. First, by shifting his representation of reptiles from killing mechanisms into sympathetic living creatures, Burdett dramatically increases the scope of the novel's central crime by adding a slew of new victims for whom he can seek justice. Further, though, in terms of detection, this shift in understanding regarding reptile lives represents a broader transformation in Jitpleecheep's Buddhist detection methodology. Karma is central to Jitpleecheep's process as a detective; his limited understanding of reptiles at the text's beginning, which indicates his inability to consider fully karmic consequences, impedes his ability to solve the crime and seek justice. Burdett frames Jitpleecheep's ineptitude as a failing not of investigative prowess but rather of ethics. That is, Jitpleecheep's failure to comprehend reptiles as sentient beings demonstrates his incomplete grasp of Buddhist morality. Since Buddhism serves as Jitpleecheep's primary detective methodology, he requires a fully formed sense of Buddhist ethics, including his ethical responsibility toward human and non-human animals, in order to detect effectively. Through transforming reptiles from weapons to victims, then, Burdett establishes animal ethics as foundational to Jitpleecheep's Buddhist detection. In this way, Burdett offers an example of detective fiction extending one of its primary generic drives, seeking justice for victims, to non-human animals as well. At the same time, Burdett's status as a British author creating a Thai Buddhist detective raises its own set of ethical questions. Ultimately, Burdett's text proves a useful exemplar for considering the ethical problematics of not only detective fiction's engagement with animals but also Western literary and scholarly works' engagement with Buddhism more broadly.

Ethics and Orientalism

In undertaking an analysis of animality in the work of a Western author writing a fictional Thai Buddhist detective, it is crucial to consider the history of Western disciplines, including animal studies, and their ethically fraught interactions with Buddhism. While Burdett uses Buddhist teachings on animal lives to imbue his detective with the ethics necessary for him to function, both Burdett's work and the discipline of animal studies are also embroiled in a larger critical debate regarding the ethics of appropriating Buddhism as an object of study in the Western academy. Donald Lopez charts the history of Western scholarly and creative engagement with Buddhism as one that, particularly regarding "the academic study of

Buddhism in Europe and America [exists] within the context of the ideologies of empire".[2] He further indicates the "Buddhist Studies, then, began as a latecomer to Romantic Orientalism ... it was in this milieu that 'Buddhism' was created in Europe" (2). Eve Sedgwick comments that, based on the history of Western interaction with and appropriation of Buddhism, "the dominant scholarly topos, and indeed, often the self-description, for Western popularizations of Buddhist thought is *adaptation*, whether such adaptation is hailed or decried".[3] By setting his series in Thailand and creating a fictional Thai Buddhist detective, the British Burdett participates in the practice of adaptation because, as Sedgwick insists, "by now a Buddhist encounter with 'Western culture' must also be understood as an encounter with a palimpsest of Asian currents and influences (and vice versa)" (157). In addition to the risk of Orientalizing that Burdett undertakes by crafting a Thai detective from a Western viewpoint, he also inescapably participates in, as Donovan Schaefer puts it, the "conversations, observations, mutations, accidents of communication, optimistic distortions, cynical manipulations, [and] bloody colonial encounters" that constitute Western interaction with Buddhist thought.[4] *Bangkok 8* and the entire *Bangkok* series provide examples of both "optimistic distortions" and "cynical manipulations";[5] in Schaefer's framing, these options are not mutually exclusive. Regardless, as Lopez insists, with Western encounters with Buddhism, be they creative, as is the case with Burdett, or scholarly, as is the case with the discipline of animal studies and, indeed, with this chapter, the complicated history of such encounters renders it nigh impossible to "offer 'alternatives,' to look forward proleptically, to assume that there can be a post-Orientalist Buddhology" (21).

Like *Bangkok 8*, the discipline of animal studies draws on Buddhism's focus on karma and its relevance to the ethical treatment of animals. Of course, Buddhism is a diverse religion with many sects, making it difficult to establish universal claims about its belief systems. Paul Waldau points out, though, that "Attitudes toward other animals, however, may be one of the few areas where generalizations are possible".[6] At the same time, Waldau also cautions that, due to the complexities of Buddhist views on animals, "inquiries about religion and animals need to be more than superficial generalizations".[7] It is worth noting here that Burdett's depictions of animals and Buddhism in *Bangkok 8* rarely make mention of the specific Buddhist texts which reference animals, of which there are many.[8] As such, Burdett's depictions run the risk of devolving into the "superficial generalizations" Waldau references. Burdett is not alone in potentially misunderstanding or misusing Buddhist perspectives on animals, however. A recent wave of interest in Buddhism from animal studies scholars has produced a similar risk of Schaefer's "optimistic distortions" or Waldau's "superficial generalizations". Dawne McCance notices that "at least to some extent, an idealized version of Eastern religions has made its way into animal studies, evident in statements as to the preference of Eastern over Western traditions, in references to the compassion for all sentient beings that lies at the core of a tradition such as Buddhism".[9] While an effort to elevate Eastern traditions over Western may avoid the "cynical manipulations" of Schaefer's observations, McCance nevertheless cautions that the "writing that emerged from this period may well be an instance of what Edward Said described in his 1979 book *Orientalism* as the kind of research that imposes a romanticized, colonialist agenda on 'Eastern' traditions" (109). Again, McCance's insight that this recent wave of interest tends to romanticize the East in general and Buddhism in particular can certainly be applied as well to Burdett's novels. *Bangkok 8*'s conclusion, which sees Jitpleecheep's FBI partner, Kimberly Jones, deciding to return to Bangkok on an "'unpaid sabbatical'" (314) exemplifies this tendency. When Jitpleecheep asks her why she has chosen to return, she replies, "'White men aren't the only ones who find this city irresistible'" (314). The American Jones' compulsion to return to an "irresistible" Bangkok serves as one example of Burdett representing the exoticized nature of Bangkok itself from a Western character's perspective. So, too, with Burdett's depictions of Buddhism. Ironically, in the case of both the animal studies scholarship McCance references and Burdett's fiction, the appeals to Buddhist traditions work to valorize Buddhist ethics yet, by their very nature, these appeals may impose "a romanticized, colonialist agenda" not at all in keeping with those same ethics. In the case of the *Bangkok* series it is important to remember the distinctions between fictional representations of religious practice and actual religious tradition.[10] Burdett's novels are, in the most generous reading, an attempt at the former: an expressly fictional representation of a culture and religious tradition distinct from the author's own.[11] Burdett's text further complicates these issues while serving as test case for the problematics of animal studies and its varied approach to Buddhist thinking, through its ambiguous description of Buddhist understanding of animality. *Bangkok 8* foregrounds the issue of whether Buddhist beliefs about animals enforce an anthropocentric hierarchical structure.

Detection, Animal Reincarnation, Karma

The complicated karmic relationship between human and non-human animals is evident in the way the text depicts reincarnation, which governs many of Jitpleecheep's actions. Buddhism's First Precept, do not kill, stems in part from a belief in reincarnation; as a Buddhist detective, Jitpleecheep is frequently torn between the seeming need to engage in violent acts in order to prevent crime and the ethics dictated by the First Precept. Waldau writes that a "reason given for the prohibition on killing is the recurring statement that all other beings have at one time been one's own father or mother. This belief is reliant upon … samsara (the different stages of reincarnation)".[12] The ban on taking life extends to animals as well, partly because, as Lisa Kemmerer suggests, the Buddhist belief in "Eons of reincarnation (transmigration) has had a predictable yet remarkable affect [sic] on the Buddhist understanding of relations: Today's duck and dog are yesterday's human lovers, siblings, and best friends".[13] Her comment is particularly relevant for Jitpleecheep, who loses his best friend and police partner Pichai in the opening pages of *Bangkok 8*. He is thus reminded that any animals he subsequently encounters in the text have the possibility to be his loved ones—indeed, throughout the text, Jitpleecheep frequently sees former animal incarnations in those around him. In describing his mentor, Jipleecheep remarks, "Now I can see the reptile in him: loose-skinned, prehistoric, cunning" (303). Jitpleecheep's sense of the former animal incarnations of those around him indicates how seriously he takes the most frequently mentioned Buddhist concept in the Bangkok series: his karma.

Jitpleecheep's karmic focus provides the framework for the ways in which his views on the ethics of reptiles transform across the course of the novel. Karma, which literally means "actions", is a relatively straightforward concept that has complex ramifications for crime as well as animals in Burdett's text.[14] Matthew Walton explains that "Buddhists believe that all of our actions have consequences that affect our future. There can be no action without a prior cause, and no action takes place without a corresponding effect".[15] Karma provides ample incentive for treating both human and non-human animals well, since the two groups cycle into and out of one another. When Jitpleecheep considers his actions, he thinks about how they will impact not only the investigation but his karma in this life as well as in his next incarnation, be it human or animal. As described in Burdett's *Vulture Peak* (2010), crime and karma are interrelated because crimes are comprehensible only in "reference to the law of karma: cause and effect. I kick you, you kick me back" (12). Thus, karma also provides a link between animals and criminality; the actions of both criminals and detectives determine whether they will be reincarnated as humans or animals. As such, Jitpleecheep's sense of ethics develops not out of altruism but rather out of self-interest. Ultimately, as Kemmerer succinctly puts it, "Karma is a law of the universe, like gravity. No one can avoid the affects [sic] of gravity or karma, whether the endangered black buck or eastern sarus crane" (101). Karma governs human and animal alike, and functions as a master concept for Jitpleecheep's life.

Bangkok 8 introduces the karmic aspects of animal reincarnation in the novel's first chapter. In a conversation Jitpleecheep has about a mob boss with his soon-to-be dead partner Sonchai, the pair muses, crudely, about the ramifications of the mob boss' behavior: "'He will be reborn as a louse in the anus of a dog.' 'Not before he's spent eighty-two thousand years as a hungry ghost'" (8). Here, the workings of karma are discussed more explicitly, and Burdett makes it plain that any mention Jitpleecheep makes to karma contains an embedded reference to animals. Upon reincarnation, the cycle of samsara continues the loop of death, life, and rebirth in one of six different realms, including the animal realm.[16] Ian Harris calls the *samsara* system "a vast unsupervised recycling plant, in which unstable but sentient entities circulate from one form of existence to the next".[17] Burdett's quotation illustrates a point of contention regarding the function of *samsara* in animal studies; tying the man's actions to his next incarnation, clearly meant as a curse—"a louse in the anus of a dog"—seems to instantiate a hierarchical system in which the man, reborn as an animal, is now of lower status than he was when he inhabited a human incarnation. In this quotation, Jitpleecheep and Pichai clearly indicate a hierarchical perspective on *samsara* rather than recognizing the sentience of the ghost or louse. Within animal studies, there is a good degree of scholarly disagreement over whether the hierarchy of *samsara* inherently makes animals inferior to humans in Buddhism. Harris suggests that *samsara* means all beings are in close kinship with one another regardless of their current incarnations (209). Waldau indicates that "Since each being is currently reaping the results of actions in past lives, the fact that a being is now a nonhuman animal, as opposed to its now being a human, suggests that it is at a lower level *by virtue of its own decisions and ways of living in the past*".[18] As a result, "To the Buddhist mind, then, in a most fundamental way, other animals' existence *must* be unhappy" and inherently inferior.[19]

Barbara Allen observes the relative nature of the hierarchy—as the Burdett citation above exemplifies, being born as an animal is not definitively a negative, because "For a hell being or a hungry ghost to be reborn in the animal realm would be good fortune and the result of positive karma" (375). As a result, she writes, among sentient beings, "all are equal and all are equally capable, over many lifetimes, of gaining enlightenment" (373). While scholars are divided on the extent to which the cycle of *samsara* places human above non-human animal lives, the quoted passage certainly shows Jitpleecheep as participating in hierarchical thinking that casts human as superior to animal. It is this perspective that shifts over the course of the novel, allowing Jitpleecheep to attain a level of Buddhist compassion that enables him to solve crimes based on empathy, insight, and karmic complexity. As he muses in *Bangkok Haunts*, "Sometimes I envy my Western counterparts the simplicity of their lives; presumably they have no care in the world beyond bringing perps to justice? A little schoolboyish, though, and lacking in moral challenge. I doubt you can burn much karma that way" (158). For Jitpleecheep, solving crimes is a means of meditating on moral complexity; any justice he brings is in service of his preeminent concern, balancing karma.

Burdett frequently links karma, and its implications for reincarnation, to Jitpleecheep's meditative detection practices. On the way to an early crime scene, for example, Jitpleecheep notes,

> I try to practice the insight meditation I learned long ago in my teens and have practiced intermittently ever since. The trick is to catch the aggregates as they speed through the mind without grasping them. Every thought is a hook, and if we can only avoid these hooks we might achieve nirvana in one or two lifetimes, instead of this endless torture of incarnation after incarnation (6).

Here Jitpleecheep positions meditation as useful for an investigative practice and links meditation to karma via reincarnation. Jitpleecheep consistently emphasizes the link between detection, meditation, and animals, saying "[A]gain I try to explain why meditation can help in the art of detection. ... 'To understand why someone suffers a violent death, it can be helpful to investigate their past lives'" (165). These past lives, of course, may include animal incarnations. Jitpleecheep brags that he actively seeks ethically complex situations in order to undergo a "moral challenge" that will strengthen his understanding of karma. *Bangkok 8* depicts just such a challenge in Jitpleecheep's interactions with reptiles in the text. However, despite being fully aware of the importance reincarnation and meditation plays in his detection and of the position that animals occupy in reincarnation, he jeopardizes both his faith and his status as detective when confronted with the *Naja siamensis* of the novel's opening sentence.

Bangkok 8 and Reptiles

Non-human animals have been integral to detective fiction since its inception, appearing as murder weapons, significant clues that precipitate the crime's solution, or even as the detective.[20] *Bangkok 8* participates in this tradition by establishing reptiles as the means by which Jitpleecheep attains a higher degree of skill in detection and transforms his thinking from an anthropocentric hierarchy into a broader need to seek justice for human and non-human animals alike. Burdett's depiction of reptiles in the early pages of the text oscillates between modes of empathetic appreciation and abject revulsion, juxtaposing Jitpleecheep's more abstract understanding of cobras from a Thai and Buddhist perspective with his more viscerally felt first-hand interactions with them. After introducing snakes as lethal in the opening lines, Jitpleecheep changes tack and explains:

> *Naja siamensis* is the most magnificent of our spitting cobras and might be our national mascot, for its qualities of beauty, charm, stealth, and lethal bite. *Naja*, by the way, is from the Sanskrit, and a reference to the great Naja spirit of the earth who protected our Lord Buddha during a dreadful storm in the forest where he meditated (4).

Here, in a passage replete with admiring language, Jitpleecheep describes three key elements of the *naja* or, as it more commonly spelled, the *naga*: its positive attributes, its dangers, and its relationship with the Buddha. These traits are central to the Buddhist understanding of the *naga*. While the term *naga* does not exclusively signify snakes,[21] "*Naga*s are often described or visually depicted as snake-like, in a hybrid form containing anthropoid and zoomorphic characteristics, or in completely human form, and wearing crowns (if described as kings)".[22] The anthropoid aspect of the *naga* present in the Buddhist tradition is of particular interest since it emphasizes the human traits which Jitpleecheep will later see in the snakes themselves. Unlike in the Judeo-Christian tradition, in which snakes frequently connote evil, deception, and temptation, *naga*s play an ambiguous role in Buddhism, making them difficult to classify.[23] Jitpleecheep's description, emphasizing both the majesty and the lethality of the cobras highlights their duality in the Buddhist tradition. The story Jitpleecheep mentions, too, is an important one in Buddhism, highlighting the significance of animals in the Buddha's early life. While Burdett does not cite the *Jataka* directly, when Jitpleecheep invokes the "great Naja spirit", he clearly references the Muccalinda king from the *Jataka* stories. Both Ivette Vargas and Barbara Allen mention this tale as an example of snakes proving beneficent in the Buddhist tradition if they are appreciated. Vargas writes,

> *Nagas* will provide stability, wealth, and wellbeing, and they will work in conjunction with the Buddhist tradition as long as they are respected for who they are and recognized for their place in the cosmos. For example, a *naga* appears early on in the biography of the historical Buddha in the role of the *Naga* King Muccalinda. This serpent shelters the future Buddha from the rain (coiling himself around the Buddha seven times) as he meditates prior to his full enlightenment experience in Bodhgaya. This tree is the only place that the future Buddha could obtain enlightenment (223).

Allen adds that "The Buddha's appreciation for the *naga* is visible; the marks on the head of the monocled cobra are said to have been made by the Buddha's fingers after he had blessed it" (403). Jitpleecheep's referencing of this particular *Jataka* tale indicates his knowledge of and appreciation for *naga*s as potentially benevolent forces in the Buddhist tradition, protectors of and marked by the Buddha himself. In an abstract way, he shares the "Buddha's appreciation for the *naga*" and does not seem to value reptiles less than human beings. His appreciation will be forgotten in an instant as the novel's first chapter presents *nagas* not as protectors but as weapons.

Bangkok 8 begins its introduction of snakes as weapons by treating the snakes as horrifying, dangerous threats that must be immediately stopped. In doing so, Burdett draws on detective fiction's precedent of using snakes to murder established by Conan Doyle in "The Speckled Band". This pivot to snakes as weapons ignores the karmic undertones of animal lives Burdett explored in the novel's previous pages and highlights Waldau's sense of *samsara* as instantiating a hierarchy between human and animal lives. Jitpleecheep and Pichai come upon the American Marine in the grey Mercedes that the novel's first sentence promised; through the window they see his head being swallowed by a giant python. The passage begins with Pichai shooting a python off of the Marine's face only to himself end up pinned beneath the Marine and the python.

> I assume that his scream is simply from fright as I go to him, at first not seeing (not believing) the small cobra which has attached itself passionately to his left eye. … Another characteristic of *Naja siamensis* is that it never lets go. I shoot it through the throat and it is only then that I understand what Pichai is trying to explain through his agony. There are dozens of them, a virtual cascade, shivering strangely and spitting as they pour out of the car. … "Don't let them reach the people. Shoot them. They must have been drugged to shake like that." He is telling me he is as good as dead, that there is no point radioing for help… No one survives a cobra bite in the eye. Already it is the size of a golf ball and about to pop, and the snakes are approaching in a narcotic frenzy (10).

Though Jitpleecheep describes the *Naja siamensis* just pages earlier in majestic terms, his actions when confronted with the cobras mark a dramatic difference. The passage emphasizes the deadly potential of the cobras as well as the giant python, and Jitpleecheep and Pichai's own use of lethal force in response. The gruesome scene is an ophidiophobe's worst nightmare, complete with a python attempting to swallow a man whole, a torrent of cobras, the horrifying behavior of the snakes shaking and spitting as if drugged (which, Jitpleecheep later confirms, they were), and the sensational image of the cobra affixed to Pichai's eyeball. The snakes demonstrate the potency of the deadly bite originally described by Jitpleecheep, rapidly taking the lives of the Marine and Pichai. The outré details such as the near-popping eyeball and the "narcotic frenzy" of the snakes create the tone of visceral disgust and horror necessary for Jitpleecheep to abandon his ethical responsibility to the snakes. These details function to distance Jitpleecheep from compassionate thinking and move him closer to considering reciprocal violence.

The scene depicts a battle of snakes versus humans, with the python having already killed the Marine and with the cobras actively killing Pichai. The scene echoes the debate in animal studies about whether *samsara* represents an anthropocentric hierarchy. Both Jitpleecheep and Pichai approach the scene with violence first, as Pichai begins by shooting the python off the Marine even though the snake poses no immediate danger to either Pichai or Jitpleecheep. Indeed, at multiple points the passage depicts Jitpleecheep and Pichai valuing the lives of humans over the lives of the snakes. Pichai's explicit concern is for the gathering crowd of spectators rather than for the cobras that he recognizes must be drugged, as his final words instruct Jitpleecheep not to "'let them reach the people. Shoot them'". Despite being beyond saving himself, Pichai directly values human lives over non-human lives here, including a directive to kill the snakes and save the humans as his last words. Beyond indicating a preference for human life over animal, Burdett depicts both Jitpleecheep and Pichai as vengeful killers, as each kills a snake for killing's sake rather than to save lives. Pichai's careful aim results in such a powerful shot that "three-quarters of the python's head is blown away" (10) and Jitpleecheep shoots the cobra attached to Pichai "through the throat". Given their violent descriptions and the fact that both the Marine and Pichai are beyond saving ("he is as good as dead … no one survives a cobra bite in the eye"), both of these graphic killings suggest the need of the detectives to revenge themselves on the snakes for the taking of human lives. Jitpleecheep in this moment chooses not only to adhere to a human–animal hierarchy but also to abandon his ethical responsibility to the living snakes.

The passage's conclusion, more than any other moment, shows how Jitpleecheep's sorrow over losing Pichai transforms him from meditative Buddhist detective to unthinking, bloodthirsty killer:

> Numb at that moment, I start to shoot them, becoming frenzied myself. I rush to the Toyota for more ammunition and change clips perhaps as many as seven times. With anguish contorting my features I lie in wait for the snakes which are trapped in the black man's clothes. One by one they writhe out of him and I shoot them on the ground. I am still shooting long after all the snakes are dead (10).

Jitpleecheep is "numb" here, "frenzied", without his meditation or flashes of insight. In these moments, full of "anguish", he neglects to think of his karma. Matthew Walton points out that, regardless of circumstance, "as with any religious tradition, it's important to note the wide gap that often exists between religious doctrine and religious practice" (74). That gap widens considerably as Jitpleecheep "rush[es] … for more ammunition" and picks the snakes off "one by one". Jitpleecheep himself introduces the *Jataka* story of the *naga* guarding the Buddha; he knows that snakes are reincarnations of his former friends and loved ones or perhaps even Buddhas-to-be, and he knows that *naga*s play a special role in Buddhist tradition. In this moment, though, under duress, in a state of numbness, he reacts by shooting "perhaps as many as seven" clips even "long after" Pichai, the Marine, and the snakes themselves are dead. Katherine Willis Perlo notes that Buddhists do have certain situations in which killing may be justifiable (122–29), but in this instance Jitpleecheep, though he is following Pichai's instructions, turns into a sniper as he lies "in wait for the snakes" and shoots past the point that "all the snakes are dead". His actions are, quite literally, overkill. Burdett, in this passage, places his detective far away from his Buddhist ideals and instantiates a speciesist hierarchy that must be corrected in order to transform Jitpleecheep into a more fully Buddhist detective. In depicting this transformation, too, Burdett implicitly argues for an alternate view of *samsara*, one in which the differing cycles of reincarnation value human and animal lives equally.

Enlightened Detection

Jitpleecheep, for all his grounding in Buddhist thought, still falters when faced with an extreme challenge such as the quiver of cobras that killed his partner Pichai. Burdett delays his redemption until the novel's final third, depicting Jitpleecheep as struggling to make headway in a difficult case without his typical degree of meditative insight. However, as he and his replacement for Pichai, the American FBI agent Kimberly Jones, go to an animal expert to explain the strange behavior of the snakes, Burdett creates a significant shift in the text. Through the intervention of an animal expert, Jitpleecheep comes to reconceive of the animals he killed not as weapons but as victims, thus reestablishing his meditative skill and restoring his detecting prowess through reconfiguring his thinking around ethical treatment of animals and its karmic ramifications.

The pivotal chapter begins with a pit full of crocodiles. As with the *naga*s, crocodiles may connote certain traits; as Mario Ortiz Robles notices, readers are trained to slot animals into certain generic tropes,[24] but in the Buddhist tradition various animals play multiple roles and are not so easily categorized.[25] In the following passages, Burdett may take advantage of the different valences available to animals in the Buddhist tradition. An ongoing question in these final passages is whether Burdett does, in fact, appeal to a multivalent understanding of non-human animals as present in the *Jataka* stories wherein reptiles can possess human anatomical traits or even, per Vargas, a "hybrid form containing anthropoid and zoomorphic characteristics, or…completely human form".[26] Given the shared Buddhist belief system between Jitpleecheep and the animal pathologist, this first explanation seems likely. Also possible, though, is a more straightforward appeal to a human–non-human hierarchy that imparts more value to reptiles the more closely they resemble humans. While the text enables either reading, Burdett indisputably softens the reptile descriptions through the words of the pathologist Trakit. In describing the crocodiles, for example, she explains that "They're very sensitive. If we make too much noise, they panic, and when they panic they pile up on top of each other, especially the younger ones, and the crocs underneath suffocate. They suffer from depression too" (167-68). Here Jitpleecheep and Jones learn in quick succession that, despite Jones' revulsion, crocodiles are very sensitive, that they are easily suffocated, and that they experience a sophisticated and relatable emotion, depression. These reptiles have quickly begun to move away from what Jones expects. If, as Norm Phelps asserts, "Buddhist morality...is based on sentience",[27] then the crocodiles have demonstrated their sentience through their capacity to feel both physical and emotional sensations. Jones expresses incredulity at the pathologist's claims while maintaining a strong aversion to the crocodiles: "Jones is almost walking on tiptoe. '*Depression?*'" (168) Throughout the passage, Jones evinces a "terminal horror" of reptiles, serving as a foil to the transformation which Jitpleecheep begins to undergo. Jones' reluctance to believe that the reptiles might be thought of as experiencing emotions parallels the callousness that Jitpleecheep displayed while gunning down the cobras. Jitpleecheep, on the other hand, chastened from his earlier experience and mindful of his Buddhist beliefs, starts to become receptive to the pathologist's perspective, evincing a Buddhist perspective divorced from a human–animal hierarchy.

The pathologist leaves the crocodiles and turns to the cobras Jitpleecheep shot, the same cobras which traumatized Jitpleecheep to the point that he says, "The problem is that suddenly I cannot look at a man without seeing a cobra gnawing at his left eye" (21). These cobras, too, will be transformed in Jitpleecheep's perspective thanks to the insight of the pathologist. She opens a drawer to reveal

> [c]obra corpses. Some of them have been neatly dissected, others are whole except for the bullet holes. "They all died of gunshot wounds, of course." She glares at me. … "they had all been poisoned with methamphetamine—

> *yaa baa*." She gives the FBI a look of the utmost sincerity. "Very few reptiles are naturally aggressive, except when hungry or protecting their young. The whole of the animal and reptile kingdom has learned to fear us, they will never attack humans unless panicked, or in this case drugged." ... "Was it laced?" "With fertilizer." Trakit shudders. "I can't think of anything more cruel" (169).

From the first, the passage anthropomorphizes the snakes, utilizing the specific genre conventions of the autopsy scene in detective fiction, which is typically a space reserved for investigating human deaths. The snakes, described as "cobra corpses" rather than carcasses, receive neat dissections and a declarative cause of death from the pathologist. These descriptions may suggest a human–animal hierarchy, with the cobras receiving greater status in Jitpleecheep's estimation due to Trakit's anthropomorphizing, while at the same time recalling the human aspects of the cobras present in the *Jataka* stories. On the other hand, Trakit's treatment of the snakes may simply indicate that they, too, are deserving of the same care and attention afforded to human victims. With the explanation of the methamphetamine laced with fertilizer, the pathologist explains both that the snakes were acting out of character when they attacked the marine and Pichai, and that this uncharacteristic behavior was precipitated by human intervention. Because of the cruelty of humans, per the pathologist's reading, reptiles have learned to fear humans, and in this case that fear was entirely justified—humans drugged them with fertilizer-laced methamphetamine. The tables have turned; when the pathologist says "I can't think of anything more cruel", she aligns sympathies with the snakes, formerly cast as nightmarish killers, and recasts humans as the cruel ones. The trope of realigning sympathy with a killer due to extenuating circumstances is a common one in detective fiction, but Burdett complicates this device through a reversal in the detective's role as well. The glare that Trakit directs at Jitpleecheep, who of course shot all the snakes in a moment of blind panic, chastens him and precipitates an epiphany due not only to the sympathy he realizes the snakes deserve but also, as a consequence, his own recasting as a villain. In these moments, the animal pathologist's treatment of the snakes works to realign the human and animal as interrelated identities per a non-hierarchical reading of Buddhist tradition. Burdett's choice of the snakes utilizes both aspects of the *naga*, lethal and beneficent, tying in the positive associations with snakes that begin the novel. The cobras have been humanized, and Jitpleecheep begins to realize that the victims of the shootout number far more than Pichai and the Marine.

The python, too, receives similar treatment, with another reconfiguration of a classic autopsy scene. The snake is kept in a drawer, just as a human corpse would be, and indeed the parallel is made explicit. Jitpleecheep comments, "'That's a very big snake.' I look at Jones, then back to the snake. It takes up the whole of a drawer that could easily fit a human being" (170). The examination proceeds, fittingly, just as a human autopsy might. The pathologist notes,

> "A mystery, the most vicious I've ever come across.' ...This drawer is huge, very deep and runs on wheels with a rumble. The python is curled up in several elongated spirals, one third of its head missing. 'He was a beauty, about ten years old, a reticulated python five meters twenty-one centimeters long.' A glance at Jones. 'That's just over seventeen feet. See the splotchy way he's camouflaged? He's native to most of Southeast Asia. Funnily enough, he lives in cities as often as the jungle. He loves riverbanks. They're an endangered species, mostly because of the illegal skin trade with China, and also for food—the Chinese love them in soup. Feel the power that must have been in those muscles" (169–70).

Several aspects of this autopsy scene work to anthropomorphize the python. The pathologist's description of the snake, complete with an opening of the drawer in the morgue to reveal the body, evokes the recounting of a murder victim's vital statistics to a homicide detective. Characterizing the crime as "vicious" and using the pronoun "he", the pathologist describes the victim's vital statistics including age, length, and fitness level, as well as typical locations. Again, these moments may be read either as anthropocentrism or as the snakes deservedly receiving the same treatment as humans. The reversal of sympathies occurs here as well, as the python has much to fear from humans, rather than the other way around, not only because the encounter with the gunman has left "one third of its head missing" but also because pythons are being illegally hunted to the point of endangerment for skin and meat. As the "glance at Jones" indicates, the pathologist moves to make her case towards the sceptical sidekick at this point, having already chastened and convinced Jitpleecheep with her earlier glare. The pathologist continues to make the karmic consequences of the snake's treatment known, remarking, "Poor poor thing. ... The drug would have induced an intense thirst, and its nerves were all on fire. Similar to being thrown into a vat of acid" (170). While the pathologist does use the word "thing" to describe the python, the expression "poor thing" is of course more typically applied to beings one cares about. The description of the pain the snake would have gone through, with its "nerves all on fire", comes with an invitation to sympathize, imagining what it would be like to be "thrown into a vat of acid". The python, which was last described swallowing a man's head, is now the object of great pity and sympathy, a creature whose time in the pathologist's lab mirrors the description of a human corpse. As with Burdett's anthropomorphizing of the cobras, it is unclear whether Trakit's foregrounding of the snake's humanoid traits marks an adherence to a hierarchy which presupposes the superiority of human animals over non-human animals. It may be, as well, that Trakit appeals to the descriptions of snakes with human anatomical features that Allen and Vargas describe as appearing in the *Jataka* stories as a way to tether her description to the Buddhist principles that she and Jitpleecheep share. This ambiguity in the text aligns with the ambiguity present in the way animal studies scholars think about karma: In this reading, either a hierarchy or an equal valuation of animal lives is possible. Regardless of whether that valuation comes from the sentience of the reptiles or their anthropomorphization, it represents the mechanism by which Jitpleecheep transforms his thinking surrounding justice for animal victims.

As the scene goes on, both Jitpleecheep and Jones become markedly more comfortable with the snakes, with Jitpleecheep picking up the tail and Jones actually touching it instead of hugging the wall in fear: "I heft the python's iron tail and gesture to Jones, who leans forward from the hips and gives it a single tentative poke with an index finger" (170). While the far more reluctant Jones is capable of only a "single tentative poke", Jitpleecheep has been receptive to Trakit's recontextualizing of the reptiles from the crocodile to the cobra to, finally, the python. As Jitpleecheep feels the weight of python's tail, he also experiences the burden of the results of his actions. As Waldau writes, "Buddhists see this orientation to the suffering of others as a *sine qua non* of ethical life".[28] The victims of the novel's initial crime now number far more than simply the Marine and Pichai, including the python and cobras autopsied so neatly by the pathologist. Not only is Jitpleecheep forced to reflect on the karmic consequences of killing the snakes, but he can also see that the crime has a far higher body count than he initially supposed. By attuning himself to the suffering of the reptiles, for which he is in large part responsible,

Jitpleecheep masters the ethical challenge the situation poses and is prepared to transform his detection to value animal lives on par with those of humans.

Conclusions

After Trakit's recontextualization of reptiles, Jitpleecheep undergoes a marked transformation in the novel's final pages. Having come close to losing his hold on the case completely, he now has a renewed sense of purpose and a new set of victims to avenge. Though of course Jitpleecheep and Pichai ended the lives of the snakes, Trakit's description of the pain inflicted on them before death also apportions a great deal of blame onto the smugglers who administered the fertilizer-laced methamphetamine. Drugging the snakes transformed them from sentient beings figured as protectors of the Buddha into vicious killers; Jitpleecheep now reverses that transformation, changing from a man who mercilessly shot dozens of snakes with no regard for their sentience or the karmic consequences of his actions into an even more meditative detective concerned with, if not protecting, at least seeking justice for his victims and rebalancing his karma. Burdett posits, through this transformation, that detectives require an ethics that renders justice incomplete if it applies only to human victims. After this transformation, with his meditative prowess restored, Jitpleecheep solves the novel's central crime in short order. Through highlighting the worth of reptiles, then, Burdett has also suggested a broader context for a master concept of detective fiction.

Ultimately, *Bangkok 8* presents a complex exemplar for thinking through the problematics of not only one of detective fiction's central tropes, the ethical application of justice, but also the ethics inherent in a Western academic and literary engagement with Buddhism. Whether this engagement inherently draws on Orientalist practices is not in question; as Lopez points out above, such engagement is predicated on a history of Orientalizing. Burdett's novel, though, reflects a tension inherent in animal studies scholarship regarding the degree to which the concept of reincarnation does or does not perpetuate an anthropomorphic hierarchy. Burdett's text, as demonstrated above, can be read as dismantling or reinstantiating that hierarchy. That the novel provides an opportunity for considering both perspectives on this problematic, as well as offering an extension on a foundational component of detective fiction, underscores the complexity of Buddhist thinking on animals, even when considered via the remove of adaptation.

Notes

1. Burdett, *Bangkok 8*, 3.
2. Lopez, "Introduction," 2.
3. Sedgwick, *Touching Feeling,* 156. Sedgwick indicates that adaptation, though it represents the dominant topos, is not the only one. She proposes *recognition/realization* as "an equally canonical topos" (156).
4. Schaefer, *Religious Affects,* 55. I am indebted to Schaefer's analysis for introducing me to the scholarly dialogue between Donald Lopez and Eve Sedgwick cited above.
5. While Burdett's intentions remain unknown, his use of Buddhism as a detective methodology, for example, might be viewed either as "optimistic distortion" or "cynical manipulation" depending on the reader's perspective. Through the former lens, Burdett's formulation of Buddhism as useful for solving crimes could stand as a well-meaning attempt to point out a practical benefit of a religious practice. Through the latter, Burdett's appropriation of Buddhist beliefs as the basis for his series might represent his exoticizing a religion as a marketing tactic.
6. Waldau, "Buddhism and Animal Rights", 83.
7. Paul Waldau, *Animal Studies,* 173.
8. The Buddhist tradition includes animals from its inception, and indeed from before its own beginnings, incorporating pre-Buddhist animal stories from India (Vargas, "Snake-Kings," 218) as well as tying animals into the story of Siddhartha Gautama's life and even birth (Allen, *Animals in Religion,* 362-63). Animals also form an important part of Buddhist didactic story tradition in the *Jatakas,* which, as Katherine Wills Perlo explains, are "tales of the Buddha's earlier lives, which on the whole encourage identification with animals by depicting the future Awakened One as having taken animal forms" (Perlo, *Kinship and Killing,* 116). In these tales, Ivette Vargas writes, animals are "used metaphorically to exemplify human values" and "appear as capable of thinking and achieving enlightened awareness" (218). Not only do the *Jatakas* present animals as embodying human traits and achieving enlightenment, demonstrating a closeness to humanity, but as Stephen Clark notes, they also show that "more literally, the Buddha was born many times in non-human form, teaching lessons of humility and generosity through his actions" (Clark, "'Ask now the beasts'", 29). In the Buddhist tradition, one reason that animal lives have value is that the Buddha himself was incarnated many times as an animal. Further, as Lisa Kemmerer points out, the past incarnations of the Buddha

point towards the possibility of future incarnations: "*Jataka* stories remind that any living being might house the karmic presence of a future Buddha; no animal is morally irrelevant" (95, italics in original). Thus, valuing animal lives matters because any non-human animal has the potential to achieve enlightenment through additional incarnations, just as any human does.

9. McCance, *Critical Animal Studies,* 109.
10. Waldau, "Buddhism", 87, parentheses mine.
11. The British Burdett lived part-time in Thailand while researching the *Bangkok* series. Fuller, "John Burdett".
12. Waldau, "Buddhism", 83.
13. Kemmerer, *Animals and World*, 101.
14. Allen, *Animals in Religion,* 374.
15. Walton, "Buddhist Reflections," 75. In concert with reincarnation, since actions impact the next life, Walton notes that "Good actions can lead to human rebirth, or in a higher realm, while bad actions can lead to rebirth as an animal, or even in a hell realm" (76). From this perspective, then, humans have a self-serving reason to practice good actions not only towards each other but also towards non-human animals; to do otherwise would risk rebirth on a lower stratum. Perlo suggests that karma provides a strong motivation to care for animals, recognizing that "Even though the unfortunate being, reborn as an animal, has erred in a previous life, if you take advantage it might be your turn in the next life" (120).
16. These six categories make up the hierarchy of *samsara*: gods, demigods, humans, animals, hungry ghosts, and hell-dwellers (Allen, *Animals in Religion*, 374).
17. Harris, "A vast unsupervised", 209.
18. Waldau, *Spectre,* 141 (italics in original).
19. Waldau, "Buddhism," 90.
20. In another, less frequent usage, Christopher Pittard discusses using animals in detective fiction to drum up sympathy for animal activism, namely antivivisectionist campaigns in Victorian England (Pittard, "Animal Voices").
21. Waldau, *Spectre,* 129.
22. Vargas, "Snake-Kings", 222.
23. As Allen points out, snakes "have often been at the centre of Buddhist traditions and their transmission. They are considered both dangerous and beneficial" (402). Ivette Vargas emphasizes the point as well, characterizing *naga*s as "Perhaps one of the most ambivalent and fascinating figures that appear throughout Buddhist literature" (222).
24. Robles, *Literature and Animal Studies*, 20.

25. Clark suggests that "there are distinctions to be made within the imagined class of 'animals'...[they] may have quite different significance in different traditions at different times" ("Animals in Religion", 571).
26. Vargas, "Snake-Kings", 222.
27. Phelps, "Buddhism and Animal Liberation", 62.
28. Waldau, *Spectre*, 96.

Works Cited

Allen, Barbara. *Animals in Religion*. London: Reaktion, 2016.

Burdett, John. *Bangkok 8*. New York: Random House, 2003.

Burdett, John. *Bangkok Haunts*. New York: Random House, 2007.

Burdett, John. *Vulture Peak*. New York: Random House, 2010.

Clark, Stephen R.L. "Animals in Religion." In *The Oxford Handbook of Animal Studies,* edited by Linda Kalof, 572-94. Oxford: Oxford University Press, 2017.

Clark, Stephen R. L. "'Ask now the beasts and they shall teach thee.'" In *Animals as Religious Subjects: Transdisciplinary Perspectives*, edited by Celia Deane-Drummond et. al., 15-34. London: Bloomsbury, 2013.

Fuller, Thomas. "John Burdett: Detective writer at work in a seedy Bangkok district." *The New York Times*, October 24, 2007.

Harris, Ian. "'A vast unsupervised recycling plant:' Animals and the Buddhist Cosmos." In *A Communion of Subjects: Animals in Religion, Science, and Ethics*, edited by Paul Waldau and Kimberly Patton, 207-217. New York: Columbia University Press, 2006.

Kemmerer, Lisa. *Animals and World Religions*. Oxford: Oxford UP, 2012.

Lopez, Donald S. "Introduction." In *Curators of the Buddha: The Study of Buddhism under Colonialism,* edited by Donald S. Lopez, 1-30. Chicago, IL: University of Chicago Press, 1995.

McCance, Dawne. *Critical Animal Studies: An Introduction*. New York: State University of New York Press, 2013.

Perlo, Katherine Wills. *Kinship and Killing*. New York: Columbia University Press, 2009.

Phelps, Norm. "Buddhism and Animal Liberation: A Family of Sentient Beings," In *Call to Compassion: Religious Perspectives on Animal Advocacy*, edited by Lisa Kemmerer and Anthony J. Nocella II, 61-72. New York, Lantern Books, 2011.

Pittard, Christopher. "Animal Voices: Catherine Louisa Pirkis' *The Experiences of Loveday Brooke, Lady Detective* and the Crimes of Animality." *Humanities* 65, no. 7 (2018), 1-16.

Robles, Mario Ortiz. *Literature and Animal Studies*. New York: Routledge, 2016.

Schaefer, Donovan O. *Religious Affects: Animality, Evolution, and Power*. Durham, NC: Duke University Press, 2015.

Sedgwick, Eve Kosofsky. *Touching Feeling: Affect, Pedagogy, Performativity.* Durham: Duke University Press, 2003.
Vargas, Ivette. "Snake-Kings, Boars' Heads, Deer Parks, Monkey Talk: Animals as Transmitters and Transformers in Indian and Tibetan Buddhist Narratives." In *A Communion of Subjects: Animals in Religion, Science, and Ethics*, edited by Paul Waldau and Kimberly Patton, 218-240. New York: Columbia University Press, 2006.
Waldau, Paul. *Animal Studies: An Introduction*. Oxford: Oxford UP, 2013.
Waldau, Paul. "Buddhism and Animal Rights." In *Contemporary Buddhist Ethics*, edited by Damien Keown, 81-112. Richmond, Surrey: Curzon Press, 2000.
Waldau, Paul. *The Spectre of Speciesism*. Oxford: Oxford University Press, 2002.
Matthew J. Walton, "Buddhist Reflections on Animal Advocacy: Intention and Liberation." In *Call to Compassion: Religious Perspectives on Animal* Advocacy, edited by Lisa Kemmerer and Anthony J. Nocella II, 73-82. New York, Lantern Books, 2011.

POLITICS

Animals, Biopolitics, and Sensation Fiction: M. E. Braddon's *Lady Audley's Secret*

Michael Parrish Lee

Why look for animals in detective fiction? A clue lies in the Victorian sensation genre that so powerfully influenced detective fiction with its plot twists and quests to uncover the bestial secrets lurking behind society's civilized facades. The opening sentence of Mary Elizabeth Braddon's iconic sensation novel *Lady Audley's Secret* (1862) puts the reader in the position of a visitor approaching Audley Court and encountering cattle that look "inquisitively at you as you passed, wondering, perhaps what you wanted; for there was no thoroughfare, and unless you were going to the Court you had no business there at all".[1] The prospect of knowing about the lives of others that Braddon's novel dangles is bound up from the start with the prospect of knowing animals and being known by them. Indeed, the kind of curiosity that we might see as the defining feature uniting the detective novel reader with the detective figure belongs first to the inquisitive cattle and then to the narrator who tries to interpret their inquisitive expressions. So, well before we get to the matter of Lady Audley's secret,

M. P. Lee (✉)
Leeds Beckett University, Leeds, UK
e-mail: M.Lee@leedsbeckett.ac.uk

R. Hawthorn, J. Miller (eds.), *Animals in Detective Fiction*, Palgrave Studies in Animals and Literature,
https://doi.org/10.1007/978-3-031-09241-1_9

the narrative raises mysteries less exclusively human: what are these cattle thinking when they look at a person, is it possible for people to know what they're thinking, and do cattle ponder similar conundrums about human thought and motives?

If these mysteries imply a ruminative pace at odds with the sensation novel's reputation for pulse-quickening plot, they are no sooner raised than, after the wink of a semicolon, abandoned with the narrator's assertion that the Court rather than this space of animals is where the business of the novel lies. I want to suggest that the novel's initial move of marking out animal life as significant in its own right only to undermine its significance structures its broader approach to "life" more generally. On one level, Braddon's narrative seems to participate in a biopolitical project similar to the production of bare life as Giorgio Agamben understands it, including animals so as to exclude them, rendering animal life disposable in order to demarcate the human life that is valuable, or at least worth reading about. Central to this dynamic is the novel's key detective figure, Robert Audley, who, we will see, is depicted as a caregiver to and non-harmer of animals. But if Robert is rarely without an animal by his side, he is also rarely far from his next mutton chop. Robert's simultaneous care for animals and complicity in their killing sheds light on his biopolitical role as a detective figure: Robert cares for the social order and for the life of his missing friend George Talboys by tracking down information about the often animalized Lady Audley that will ultimately lead to her incarceration, abandonment, and death in a Belgian *maison de santé*. By exploring relationships with animals, the novel thus tests out how human life might be simultaneously known, cared for, and abandoned. However, this essay will also show that such abandonment of animal life is never complete and must be played out repeatedly, as animals emerge again and again as objects of knowledge, subjects of care, conscious agents, and figures of abandonment. By attending to this dynamic, we not only find Braddon's novel grappling with the biopolitical entanglements of caring for life and abandoning life, but we also discover how Braddon's attention to animals ultimately opens up spaces that resist and refuse the abandonment of life that would seem to structure the book's larger narrative.

The abandonment of life haunts the biopolitics that Michel Foucault describes emerging at the end of the eighteenth century when "the biological existence of a population" became a key concern[2] and "power" took biological "life under its care".[3] This emergence shifted the emphasis of state-sanctioned killing from "victory over political adversaries" to the supposed protection of life and included forms of "indirect murder" such as "increasing the risk of death for some people".[4] Expanding on Foucault's analysis of biopolitics, Agamben argues that the Western political order demarcates itself as a worthwhile "form or way of living" by distinguishing itself from a kind of unqualified bare or "natural" life[5] that, because unqualified, can be treated as "life devoid of value" (139) and even "life that may be killed" (89). For Agamben, bare life is incorporated into the political order in the form of that which is excluded and can be expelled from this order. Along similar lines, Emily Steinlight suggests that the Victorian novel's biopolitical work of "de[aling] out life and death"[6] entailed life "being systematically valued in the very process by which certain bodies and lives came to appear disposable" (116). For Steinlight, as for Foucault, the life pertinent to biopolitics is human life. I want to suggest, however, that *Lady Audley's Secret* poses questions over life's value and disposability most fundamentally at the level of animal life and that Braddon's use of animals is crucially entwined with and essential to understanding her novel's biopolitical treatment of human life.[7]

The nineteenth century, after all, not only saw the rise of human biopolitics but also the entry of non-human animal life into the domains of knowledge and care in new ways. Harriet Ritvo notes that the "beginnings of the animal protection movement in England" can "be traced to the end of the eighteenth century".[8] The nineteenth century brought the founding of the Society for the Prevention of Cruelty to Animals in 1824, the passing of the Cruelty to Animals Acts of 1835, 1849, and 1876, and "the rise of bourgeois pet-keeping", which Keridiana W. Chez argues was "inextricable" from the "humane movement".[9] But the potential flip side of this care for animals involved new modes of commodification and marginalization. John Berger connects the popularity of pets to the nineteenth-century manufacture of "realistic animal toys", "widespread commercial diffusion of animal imagery", and establishment of modern zoos, which he sees as comparable to "sites of enforced marginalisation" such as "prisons, madhouses, concentration camps".[10] And Jacques Derrida, drawing on the work of Henri F. Ellenberger, suggests that zoos and psychiatric hospitals emerged as parallel institutions that "had the ambition or the pretension to treat, to care for, to take great care (*cura*) of what it was enclosing and objectifying and cultivating".[11]

Such nineteenth-century slippages and overlaps between, on the one hand, interest in and care for animal life, and on the other hand, domination, commodification, and marginalization of such life, mean that animals offer a vital lens through which to consider the period's biopolitics, and one that reveals that questions about animals are crucially entangled with biopolitical conceptions of human life. In *The Open*, Agamben developed his ideas about bare life by aligning it more explicitly with animal life, suggesting that it is "possible to oppose man to other living things, and at the same time to organize the [...] economy of relations between men and animals, only because something like an animal life has been separated within man".[12] In a comparable fashion, *Lady Audley's Secret* seems to incorporate animals in order to mark animal life as that which is excluded and excludable from its narrative, and to similarly mark those "animalistic" humans who have not sufficiently separated themselves from an animal life within them. Such marking helps us understand detective fiction's simultaneous reliance upon and disavowal of criminality as an animalized state of humanity that must be exposed and vanquished.

Victorian critics of sensation fiction sometimes dismissed it as an animalistic genre. For instance, an 1863 article titled "Our Female Sensation Novelists" expressed concerns that sensation fiction stimulates readers'—and particularly female readers'—attention "through the lower and more animal instincts",[13] and an 1866 review of Wilkie Collins' *Armadale* (1866) complained: "bigamy has been Miss Braddon's big black baboon, with which she has attracted all the young girls in the country. And now Mr. Wilkie Collins has set up a big black baboon on his own account".[14] Similarly, in his 1863 article, "Sensation Novels", Henry Mansel condemned "this ravenous appetite for carrion, this vulture-like instinct which smells out the newest mass of social corruption, and hurries to devour the loathsome dainty before the scent has evaporated".[15] Ann Cvetkovich points out that, "[a]ccording to his discourse, the sensation novel is deplorable because it reduces its readers to the condition of animals who are driven by instincts".[16] And Susan D. Bernstein, noting the influence of Darwinian ideas on sensation fiction and reactions to the genre (Charles Darwin's *Origin of Species* appeared in 1859), argues that such responses convey "contemporary fears of evolutionary transformation not as progress but as degeneration"[17]—hence the image of the big black baboon. But what if we go beyond acknowledging concerns over sensation fiction as an animalistic genre capable of producing animalizing affects to contend with the complexities that arise when we pay attention to the actual animals that inhabit the pages of this fiction? I will accept the Victorian critics' point about sensation fiction being an animalistic genre, but I will take the point in a more positive and literal sense: sensation fiction is a genre that tests out the relationship between humans and animals. Under its influence, detective fiction developed an obsession with criminality that we might understand as an obsession with locating, establishing, or even problematizing the threshold where "the human" meets "the animal".

If part of what Victorian critics found troublingly animalistic about sensation fiction was an association of the genre with narrative "Action, action, action!" and the resulting nervous "excitement",[18] Braddon's initial depictions of non-human animal life share neither of these qualities, instead aligning such life with stasis and tranquillity. After glimpsing cattle at the outskirts of Audley Court, we enter for a tour of the place—"a place that visitors fell into raptures with; feeling a yearning wish to have done with life, and to stay there for ever, staring into the cool fish-ponds, and counting the bubbles as the roach and carp rose to the surface of the water".[19] While the cattle at the opening of the novel emerged as life that must be passed over before arriving at the true "business" of the story, here, at the purported centre of this business, we encounter abundant animal life reconfigured as the absence of life, a pond full of animated fish recast not merely as a site of tranquillity but as a lulling pool of non-being. Two chapters later, Braddon compounds the association between animals and the absence of life:

> The lowing of a cow in the quiet meadows, the splash of a trout in the fish-pond, the last notes of a tired bird, the creaking of wagon-wheels upon the distant road, every now and then breaking the evening silence, only made the stillness of the place seem more intense. It was almost oppressive, this twilight stillness. The very repose of the place grew painful from its intensity, and you felt as if a corpse must be lying somewhere within that grey and ivy-covered pile of a building—so deathlike was the tranquillity of all around (26).

On one level, Braddon associates animals here with a kind of unnarratable or subnarratable pastoral that the sensational plot will disrupt.[20] In this respect, the corpse reference foreshadows the murderous story that will unfold. But, more than this, Braddon configures animal life as a space of non-life, or one where life is suspended. It is not just that this scene is so still that we might long for the intrusion of plot in the form of a murder, but also that the deathlike tranquillity here already is like a corpse and prefigures the appearance of any actual corpse to come. The subnarratable life that might be intruded upon by narratable murder is already positioned with death. Animal life here represents life that has not begun to qualify as life, whose status *as* life is perpetually suspended. Ivan Kreilkamp suggests that "animal characters are fundamentally 'minor,' in the sense defined by Alex Woloch"[21]—that is "subordinate", "delimited", and functional.[22] But Braddon seems to go beyond this and make animal life the vanishing point of life itself.

Braddon also uses animals to mark out valueless human life, particularly when animals become figurative or function as comparisons for humans. Robert Audley's cousin Alicia complains: "To have only one cousin in the world, [...] my nearest relation after papa, and for him to care about as much for me as he would for a dog!"[23]. This might appear an odd comparison because Robert is particularly fond of dogs and Alicia herself has an ongoing interest in the animals that surround her. She is introduced as "an excellent horsewoman" who "spent most of her time out of doors, riding about the green lanes, and sketching the cottage children, and the ploughboys, and the cattle, and all manner of animal life" (10). And her dog Cæsar is "the sole recipient of the young lady's confidences" (92). But this love of animals only brings into relief her perceived deprivation of the love that she really wants, that of her father and that of her cousin. Her very valuing of animal life becomes a way of showing how she is barred from the life of "real" value: Significance within the human social world that for a woman in her position might entail either "reign[ing] supreme in her father's house"—something she no longer does since her widowed father Sir Michael Audley has married the titular Lady Audley (10)—or entering into a conjugal life with her romantic interest and cousin Robert. The animals in her life emerge as mere consolation for the life she cannot have: "She had her favourite mare, her Newfoundland dog, and her drawing materials, and she made herself tolerably happy. She was not very happy" (249). Yet, for Alicia, animal life eventually changes from consolation into failed consolation, from better than nothing into nothing itself: "It seemed very hard to be a handsome grey-eyed heiress, with dogs and horses and servants at her command, and yet to be so much alone in the world as to know of not one friendly ear into which she might pour her sorrows" (249). Her dog Cæsar has transformed from her companion and confidante into something "at her command" that only serves as a painful reminder that she now seems to have *no* confidante at all. And, even if such command would seem to confer power, that power appears relatively empty at a point when she feels that her father has now "accepted a new ruler" in Lady Audley (249). In his vacillation between being a friend signifying an absence of friends and an object of power signifying an absence of power, Cæsar reveals animal life in *Lady Audley's Secret* as the crucial site where life converts into valueless life. Moreover, the lumping together of servants with such devalued life as worthless objects of command shows how animals are used to mark out disposable life in the human world.

Luke Marks, the husband of Lady Audley's maid Phœbe, provides an even more striking example of how animalization marks out disposable human life. Like Alicia, Luke compares his treatment to that of a dog. Late in the novel he says of Lady Audley: "Whatever she give me she throwed me as if I'd been a dog. Whenever she spoke to me, she spoke as she might have spoken to a dog" (367). But if he uses this animal comparison to show his feeling of degradation, the narrator introduces him in a way that suggests that such a comparison is not unjust, describing him as a "stupid-looking clodhopper" with a mouth that is "coarse in form and animal in expression" and adding that "he was not unlike one of the stout oxen grazing in the meadows round about the Court" (28). Where bovine life at the edges of Audley Court once, briefly, signalled inquisitive consciousness, now a bovine comparison comes to signal coarse stupidity. And while the cattle at the beginning of the novel are abandoned in narrative terms, and the fish in the fishpond are reconfigured as an absence of life, the animalized Luke Marks is ultimately ejected from the novel through his actual death. In *Lady Audley's Secret*, not only animal life, but also animalistic human life, appears disposable.

My analysis so far suggests that Braddon's treatment of animal life in narrative terms is similar to the political order's simultaneous inclusion and exclusion of bare life as described by Agamben. Yet, if Braddon uses animal life to mark out life that can be abandoned, she also makes it a privileged site of care. The novel introduces Robert Audley, like Alicia, partly through his relation to animals. Before the disappearance of his friend George Talboys (later revealed as husband to the bigamous, identity-disguising Lady Audley) rouses him to active sleuthing, Robert Audley appears as an idle, non-practising barrister, "a handsome, lazy, care-for-nothing fellow" (32). But this care for "nothing" turns out to be a care for animals: we find that Robert's barrister's chambers "were converted into a perfect dog-kennel by his habit of bringing home stray and benighted curs" (33). Robert's initial inactivity and non-participation in the practice of law seems synonymous with him opening up his legal chambers as an animal refuge. Robert also appears to be a practitioner of non-violence towards animals, someone "who would not hurt a worm" and who spends "the hunting season at Audley Court" but who "keep[s] at a very respectful distance from the hard riders; his horse knowing quite as well as he did that nothing was further from his thoughts than any desire to be in at the death" (33). Robert's much commented on non-normative or incomplete masculinity[24] initially manifests as a conglomeration of passivity, a lack of interest in the law, and a kindness to animals. He therefore seems to sidestep the "*carno-phallogocentrism*" that Derrida sees as structuring Western humanist conceptions of male subjectivity that connect *logos*—the command of language, meaning, and knowledge—to "carnivorous virility" and a willingness to sacrifice life.[25]

The problem is, Robert likes his meat. Whether eating mutton chops[26] or ham and eggs (116–17), complaining that his duck dinner is cold (73–74), or attending a "luxurious eating-house" because tired of the dishes served by Mrs. Maloney, "whose mind ran in one narrow channel of chops and steaks" (176), Robert not only consumes animal flesh, but is something of a connoisseur. Sitting down to "the familiar meal" of a mutton chop, Robert recalls "his uncle's cook with a fond, regretful sorrow", thinking "sentimentally" of how her "cutlets à la Maintenon made mutton seem more than mutton; a sublimated meat that could scarcely have grown upon any mundane sheep" (132). And in the same speech where he insists that, despite spending the hunting season at Audley Court, he doesn't "care for" shooting and "never hit a bird in [his] life", Robert remarks that he only visits his uncle's house for "the change of air, the good dinners, and the sight of [his] uncle's honest, handsome face" (48). Robert disavows animal killing while praising what are almost certainly meat meals, but his carnivorous appetite and care for animals are not merely contradictory. Rather, together they highlight the ease with which animal life is abandoned to death; even those who care most about animals are often complicit in their killing.

Robert's quasi-paradoxical treatment of animals shows how the abandonment of life is built into the structure of the biopolitical care for life. He not only feeds his own carnivorous appetites, but those of others, including animal others. Robert, for instance, sits down to breakfast "with one of his dogs at each side of his armchair, regarding him with watchful eyes and opened mouth, awaiting the expected morsel of ham or toast" (120). For Robert, caring for animals means feeding them animals. While looking after the missing George Talboys' son Georgey, Robert connects childcare with both animal care and eating animals. We learn that he "had catered for silk-worms, guinea-pigs, dormice, canary birds, and dogs, without number, during his boyhood, but he had never been called upon to provide for a person of five years old" and that his memory of his own five-year-old diet is "of getting a good deal of bread and milk and boiled mutton" (153). However, Georgey turns out to be even more of an "epicure" than Robert is, "reject[ing] milk and bread and ask[ing] for veal cutlets" (154). The irony that this child's meat of choice is the meat of an animal child is furthered when we consider that, as Kathleen Kete observes, there was a nineteenth-century imaginative "link between animals and children".[27] It is as though Georgey elevates himself from mere child to "little gentleman"[28] in an act of sacrificial sophistication that distinguishes his child's life from the animal life of the child that he eats. And Georgey ends up with a meal nearly as abundant in animal life as Robert's childhood was, eating "stewed eels, a dish of cutlets, a bird, and a pudding" (154). In *Lady Audley's Secret*, the designation of some life as bare, unqualified, disposable, killable, or edible seems necessary not only for feeding life but also for marking out the life that is not bare, the life of quality and value that is worth caring for. Earlier in the novel, Robert cares for Georgey's father at Audley Court as George grieves the (reported) death of his wife (who turns out to be the woman going by the name Lucy Graham, now Lady Audley), and this care is largely oriented around taking George fishing (49, 57–58, 69–71). Even if the purportedly non-violent Robert prefers fishing to shooting because "you've only to lie on a bank and stare at your line; I don't find that you often catch anything, but it's very pleasant" (49), the pastime relies on the status of fish as life that may legitimately be killed. Through Robert's acts of care and consumption we see the sacrificial abandonment of life to death as itself a form of care—as both self-care and care for others.

Fitting, then, that Robert's major acts of care entail transforming Lady Audley into bare life. George Talboys' disappearance spurs Robert from passive, animal loving "care-for-nothing fellow" to active detective, now focusing the majority of his care on his missing human friend. His quest for knowledge about what happened to George leads him to discover that not only is Lady Audley responsible for his disappearance but also that she is hiding her identity as his wife while married to Robert's uncle Sir Michael Audley. Robert's pursuit of Lady Audley now also becomes about protecting his beloved uncle and the patriarchal social order that Sir Michael stands for. Critics such as Pamela K. Gilbert and Steinlight have argued that the text links Lady Audley's problematic femininity to a biological or biopolitical threat. Gilbert sees Lady Audley as a figure of "women's sexuality" that is "represented as a contagious disease",[29] a dangerous open body that must be "contained" (93) with the help of Robert, the narrative's "sanitary policeman" (105). And Steinlight reads her as "an ungovernable female body" that "incorporate[s] specters of mass population".[30] Lady Audley is certainly a figure of uncontainable life, but little serious attention has been paid to how this excess life becomes synonymous with animal life. The narrator describes her feelings "t[earing] at her like some ravenous beast",[31] compares her stealthy "footfall" to "that of some graceful wild animal" (268), and characterizes the "flame" in her eyes as a light "such as might flash from the changing hued orbs of an angry mermaid" (273). And Robert dreams

> he saw Audley Court, rooted up from amidst the green pastures and shady hedgerows of Essex, standing bare and unprotected upon that desolate northern shore [...]. As the hurrying waves rolled nearer and nearer to the stately mansion, the sleeper saw a pale, starry face looking out of the silvery foam, and knew that it was my lady, transformed into a mermaid, beckoning his uncle to destruction (209–210).

Both Robert and the narrator imagine Lady Audley's threat in terms of animal life lurking within or intermingling with human life. While the novel's opening sentence gave us animals in their proper place, cattle grazing in enclosed land, looking over "high hedges" from meadows that "bordered" the avenue (1), Lady Audley shows animal life breaching the borders of propriety and property:[32] she embodies all at once the hungry animal that tears at the human, the wild animal that sneaks through and out of houses, and the aquatic animal that merges with the human, ripping Audley Court away from the very pastures and hedgerows that seemed to mark animal life as essentially docile and containable.

Yet if the containability of the cattle that begin the novel is part of what makes them appear so immediately disposable in a narrative sense, it is the uncontainable animal life within Lady Audley that seems to guarantee her ultimate social and narrative disposability. Upon exposing Lady Audley, Robert, despite inconclusive medical evidence of her madness, has her confined to a Belgian institution where she can be locked away without recourse to a legal trial that could ruin his uncle's good name. In the chapter titled "Buried Alive",[33] Lady Audley likens her fate to living death in a "living grave" (333). Lady Audley, in other words, comes to inhabit the position of suspended life that animals in the novel so often occupy. It is here, abandoned to a supposed institution of care (an institution, we recall, that Derrida links to the zoo), in an extraterritorial space beyond Britain at the murky threshold between law and medicine, as bare life caught between life and death, that Lady Audley ultimately dies and is expunged from the novel.

As a narrative and biopolitical solution, however, Lady Audley's incarceration and death proves insufficient. Despite the excision of Lady Audley's story from the "narrative of the healthy social body" in "order to maintain its unity and health," Gilbert suggests that the novel's ending "cannot negate the subversive insistence of the Lady's voice".[34] And Steinlight argues that removing "this supposedly abnormal specimen from the English population" can "scarcely remedy the larger problem [of surplus population] that she instantiates" and that the narrative ultimately "erodes the conventional boundary between the self-regulating individual and the unmanageable multitude".[35] The narrative problem that Lady Audley poses, in other words, is less that she is an abnormal specimen of femininity than that she points to Victorian concerns about the potential disruptive excess of womanhood in general and fears that, as Lyn Pykett puts it, "women cannot be contained within dominant definitions of 'woman', or of normal femininity".[36] Such concerns stem partly from nineteenth-century beliefs that women were "driven by their bodily processes"[37] and that they, like children, were "closer to nature" than "bourgeois men".[38]

Braddon draws upon but also parodies such associations of women with bodily and animal life through Robert Audley's exuberantly misogynist meditation on the slim chances of happiness in marriage. He wonders: "Who is to say which shall be the one judicious selection […]? Who shall decide from the first aspect of the slimy creature, which is to be the one eel out of the colossal bag of snakes?".[39] For Robert, women as potential mates are only ever "creatures", and the choice of whom to marry constitutes the choice between more and less appealing (and perhaps edible) varieties of animal. Robert's animalization of women as marital prospects (in conjunction with his foregrounding of sliminess) chimes with Kete's point that women were seen as particularly close to nature largely because of their association with reproduction.[40] Such an association brings us to a key biopolitical problem that haunts *Lady Audley's Secret* (and has repercussions well beyond the novel): Victorian women occupy a privileged biopolitical position as reproducers of the population, but in this position, tied so closely to biological bodily functions, they also occupy a place imagined as close to animal life and are thus more susceptible to slipping (or being forced) into the position of bare life that animals occupy. To put it differently, the life animalized and expelled through the figure of Lady Audley is not ultimately separate or separable from the regenerative life envisioned at the end of the novel where the renewal of the social body plays out in the reproductive marriage plot with the union of Robert and George's sister Clara bringing forth a "baby who has just begun to toddle".[41]

Throughout her narrative, Braddon draws attention to rather than merely relies on the paradox of treating life as simultaneously valuable and disposable. So while Agamben's understanding of bare life is useful up to a point in clarifying the biopolitical workings of *Lady Audley's Secret*, his model does not allow for the complexity with which Braddon's narrative poses questions about life and its treatment. If Braddon draws on a sacrificial mode throughout her narrative, she also exposes the workings of this mode and the cruel ironies inherent in it. Consider, for instance, Braddon's association of Lady Audley with aquatic animal life in conjunction with the treatment of actual aquatic life throughout the narrative. We might recall Robert's propensity for fishing and his commitment to sharing his hobby with the grieving George Talboys; the novel asks us to imagine the consciousness of fish at precisely the moment that it conjures them as disposable life: "Those were happy fish in the stream on the banks of which Mr Talboys was seated. They might have amused themselves to their heart's content with timid nibbles at this gentleman's bait, without in any manner endangering their safety; for George only stared vacantly at the water" (71). While the fish partly serve as a backdrop to George's troubled mental state, their consciousness is also a remainder that troubles the seemingly straightforward equation of animal life with bare life. In this regard, rendering someone or something animalistic in the novel might seem to legitimize their transformation into disposable life, but Braddon casts this model as an insufficient one designed to provoke unease.

Such unease amplifies when we note further connections between aquatic life and the mermaid-like Lady Audley. The bank of the stream is George's initial site of disappearance, "the fishing-rod lying on the bank" the first sign that he is missing (73) (it turns out that Lady Audley pushes him down a well). In the chapter directly following the exposure of Lady Audley's double identity and bigamy, Alicia, ever at odds with the lady, hungrily announces a fish dinner: "*Is* papa coming to dinner? [...] I'm *so* hungry; and [...] the fish will be spoiled. It must be reduced to a species of isinglass soup by this time, I should think" (306; italics in original). Finally, sentences before we learn of Lady Audley's death, we read that her son Georgey—who now, in the novel's ostensibly happy ending, makes up part of an extended family consisting of his father, Clara, Robert, and the baby—spends his time "fish[ing] for tadpoles" (379). The sacrificial position of the now-caught fishy Lady Audley is mirrored in the fate of the novel's other aquatic creatures.

The novel poses further questions about the treatment of life and the intersections between care and killing through Lady Audley's relationship with Alicia's dog Cæsar. During a meeting of Lady Audley and Alicia, "[t]he dog, which had never liked my lady, show[s] his teeth with a suppressed growl" (72). On one level, Cæsar's reaction aligns with one of the main roles of animals in melodrama, which, according to John MacNeill Miller, is to function as figures of goodness with an access to "the cosmic moral order" that aids them in "hunting down villains".[42] Cvetkovich and Pykett note the sensation novel's important roots in and adaptation of stage melodrama,[43] and, in this context, Cæsar's dislike of Lady Audley functions as an early clue; sniffing out her moral corruption, Cæsar shows an animal playing a vital role in what Patrick Brantlinger describes as the sensation novel's crucial work of "stripping away surface appearances".[44] Yet, Pykett observes that women's sensation fiction is "more conflicted and ambiguous" than stage melodrama,[45] and Lady Audley's reaction to Cæsar suggests something less straightforward than the dog occupying either a position of bare animality or one of melodramatic goodness: "Bah, Cæsar; I hate you, and you hate me; and if you met me in the dark in some narrow passage you would fly at my throat and strangle me, wouldn't you?".[46] Neither just a revelation of moral truth nor a sign of purely animalistic instinct, Cæsar's "suppressed growl" points to restraint and perhaps, as Lady Audley suggests, the all-too-human feeling of hatred. But if this hate is at once mutual between the lady and the dog and tied to bared teeth and suppressed growls, then the text also dissolves any absolute distinction between animalistic violence and more seemingly sophisticated human feelings. Lady Audley's address to Cæsar suggests that she sees herself and him, underneath their present disparity in social power, as uncannily close to being on even ground.

Sir Michael Audley's response to this interaction draws particular attention to the fine and shifting line between the care for life and the disposability of life. When Alicia tells her father that if she "had not had hold of his collar" Cæsar would indeed "have flown at [Lady Audley's] throat and strangled her", Sir Michael answers, "Your dog shall be shot [...] if his vicious temper ever endangers Lucy" (93). While dogs throughout the novel are figured as subjects of care and we know that Cæsar, in particular, is Alicia's companion, any right he has to live gets nullified in the name of protecting the life of Lady Audley. But as soon as Sir Michael utters this threat, Braddon spends a whole paragraph hinting at Cæsar's mental and emotional life:

> The Newfoundland rolled his eyes slowly round in the direction of the speaker, as if he understood every word that had been said. Lady Audley happened to enter the room at this very moment, and the animal cowered down by the side of his mistress with a suppressed growl. There was something in the manner of the dog which was, if anything, more indicative of terror than of fury, incredible as it appears that Cæsar should be frightened of so fragile a creature as Lady Audley (93).

If Lady Audley will, on the one hand, prove far from fragile in her two attempted murders in the narrative, her eventual fate, on the other hand, shows her finally having no more claim on life than her fellow creature Cæsar does. Braddon uses animal life to apparently legitimize but ultimately problematize the demarcation of disposable life akin to bare life. She does so not only by evoking the inner lives of animals, whether inquisitive cattle, happy fish, or frightened dogs, but also by frequently evoking them in the very moments that such life is threatened or marked as disposable. So, despite employing a sacrificial mode where caring for life cannot be separated from abandoning life, Braddon hints, on another level, at an ethics of non-abandonment.

What would such an ethics look like? It can only be glimpsed in the actual narrative of *Lady Audley's Secret* as a series of possible or virtual lives, a menagerie of the unnarratable. Caring for these lives might mean stopping, for instance, with the cattle that begin the novel and the fish that seem initially only to signify tranquillity; it might mean staying with Lady Audley after Robert's final visit to her until the moment of her death. D. A. Miller has memorably characterized the sensation novel-reading experience as one of page-turning, nerve-tingling "physicality",[47] a plot-driven pleasure that Cvetkovich observes led Victorian critics to fear "the prospect of a reader reduced to a body reacting instinctively to a text", to "the condition of animals".[48] But paying attention to the animals in *Lady Audley's Secret* means discovering that, for Braddon, animal life is no mere reduction but an opening of possibility disguised as life's vanishing point. Braddon ultimately suggests that animal life is not some pure, unqualified bare life or raw physical material despite its frequent treatment as such. And if Braddon's animals aren't disposable matter but life we cannot abandon, then it is harder than ever to imagine that her fiction's agenda is the reduction of readers to reactive bodies carried by propulsive plot.

Such life instead invites something closer to Miller's more recent practice of "Too Close Reading"—one "drawn to details that, while undeniably intricate, are not noticeably important"[49] and that, if attended to, halt "narrative flow" (125), derailing the plot, or at least putting it "on pause" (114). But while Frances Ferguson celebrates Miller for enabling an approach in which the objects of readerly attention are "merely and, supremely, personal",[50] Braddon's animals call readers to step beyond the personal and towards a responsibility to other lives that seem, almost emphatically, not to matter. This call for non-abandonment reveals the sensation novel harbouring a countercurrent to the very protocols it helps develop and that inform detective fiction's impulse "to restrict and localize the province of meaning"[51] and the detective figure's promise of knowledge as mastery.[52] If we can risk a reading that loses the plot, these animal lives will lead us gloriously astray, and not just to passed-over details buried in the text but to the cusp of the stories they might tell if the narrative would let them. What if one of the principal authors of a genre supposedly all about plot were drawing us all along to the lives that take us to and past plot's edges? Perhaps the supposedly animalizing speed of the sensation novel is itself a wrong turn from the animals at once included in and excluded from it, beckoning us, often with nothing more than a look, a bubble, or a roll of the eyes, to slow or stop and re-imagine disposable life as something we cannot turn away from.

Notes

1. Braddon, *Lady Audley's*, 7.
2. Foucault, *The History of Sexuality* Volume 1, 137.
3. Foucault, "*Society*", 253.
4. Foucault, "*Society*", 256.
5. Agamben, *Homo Sacer*, 1.
6. Steinlight, *Populating the Novel*, 142.

7. I thus follow in the footsteps of critics such as Nicole Shukin and Cary Wolfe who call for more dialogue between biopolitics and animal studies.
8. Ritvo, *Noble Cows*, 74.
9. Chez, *Victorian Dogs*, 2.
10. Berger, *About Looking*, 26.
11. Derrida, *The Beast and the Sovereign* Volume 1, 300.
12. Agamben, *The Open*, 15–16.
13. "Our Female Sensation Novelists", 210.
14. Review of *Armadale*, 269.
15. Mansel, "Sensation Novels", 502.
16. Cvetkovich, *Mixed Feelings*, 20.
17. Bernstein, "Ape Anxiety", 259.
18. Mansel, "Sensation Novels", 486.
19. Braddon, *Lady Audley*, 8.
20. I borrow the term "the unnarratable" from Prince, "The Disnarrated", 1, and Warhol, "Neonarrative", 221. Warhol describes the variety of the unnarratable she calls the "*subnarratable*" as that which is "too insignificant or banal to warrant representation" (222).
21. Kreilkamp, "Dying Like a Dog", 82.
22. Woloch, *The One vs. the Many*, 27.
23. Braddon, *Lady Audley*, 60.
24. See for example Taylor, *In the Secret Theatre of Home*, 11, Cvetkovich, *Mixed Feelings*, 57, Gilbert, *Disease, Desire, and the Body*, 100–101, and Pykett, *Improper Feminine*, 103–104.
25. Derrida, "'Eating Well", 280.
26. Braddon, *Lady Audley*, 207–209.
27. Kete, "Introduction", 7. See also Cosslett, "Child's Place in Nature", especially 479–80.
28. Braddon, *Lady Audley*, 154.
29. Gilbert, *Popular Novels*, 8.
30. Steinlight, *Populating the Novel*, 141.
31. Braddon, *Lady Audley*, 241.
32. See also Langland, "Enclosure Acts", who suggests a relationship between the containment of women in the text and the agricultural enclosure acts.)
33. Braddon, *Lady Audley*, 325.
34. Gilbert, *Popular Novels*, 105.
35. Steinlight, *Populating the Novel*, 147.
36. Pykett, *Improper Feminine*, 95.
37. Mangham, *Violent Women and Sensation Fiction*, 13.
38. Kete, "Introduction", 8.
39. Braddon, *Lady Audley*, 175.
40. Kete, "Introduction", 8.

41. Braddon, *Lady Audley*, 379.
42. Miller, "When Drama Went to the Dogs", 531.
43. See Cvetkovich, *Mixed Feelings*, 24 and 45, and Pykett, *Improper Feminine*, 74–76.
44. Brantlinger, *The Reading Lesson*, 144.
45. Pykett, *Improper Feminine*, 76.
46. Braddon, *Lady Audley*, 72.
47. Miller, *Novel and the Police*, 147.
48. Cvetkovich, *Mixed Feelings*, 20.
49. Miller, "Hitchcock's Hidden Pictures", 126.
50. Ferguson, "Now It's Personal", 540.
51. Miller, *Novel and the Police*, 34.
52. See, for example, Siddiqi, *Anxieties of Empire*, 19.

Works Cited

Agamben, Giorgio. *Homo Sacer: Sovereign Power and Bare Life*. Translated by Daniel Heller-Roazen. Stanford: Stanford University Press, 1998.

Agamben, Giorgio. *The Open: Man and Animal*. Translated by Kevin Attell. Stanford: Stanford University Press, 2004.

Berger, John. *About Looking*. London: Bloomsbury, 2009.

Bernstein, Susan D. "Ape Anxiety: Sensation Fiction, Evolution, and the Genre Question". *Journal of Victorian Culture* 6, no 2 (2001): 250–271.

Braddon, Mary Elizabeth. *Lady Audley's Secret*, edited by Lyn Pykett. Oxford: Oxford University Press, 2012.

Brantlinger, Patrick. *The Reading Lesson: The Threat of Mass Literacy in Nineteenth-Century British Fiction*. Bloomington and Indianapolis: Indiana University Press, 1998.

Chez, Keridiana W. *Victorian Dogs, Victorian Men: Affect and Animals in Nineteenth-Century Literature and Culture*. Columbus: Ohio State University Press, 2017.

Cosslett, Tess. "Child's Place in Nature: Talking Animals in Victorian Children's Fiction". *Nineteenth-Century Contexts* 23, no 4 (2002): 475–95.

Cvetkovich, Ann. *Mixed Feelings: Feminism, Mass Culture, and Victorian Sensationalism*. New Brunswick, NJ: Rutgers University Press, 1992.

Derrida, Jacques. "'Eating Well,' or the Calculation of the Subject". In *Points ... Interviews, 1974-1994*, translated by Peggy Kamuf et al, edited by Elisabeth Weber, 255–87. Stanford: Stanford University Press, 1995.

Derrida, Jacques. *The Beast and the Sovereign* Volume 1. Translated by Geoffrey Bennington. Edited by Michel Lisse, Marie-Louise Mallet, and Ginett Michaud. Chicago and London: University of Chicago Press, 2009.

Ferguson, Frances. "Now It's Personal: D. A. Miller and Too-Close Reading". *Critical Inquiry* 41, no. 3 (Spring 2015): 521–40.

Foucault, Michel. *"Society Must Be Defended": Lectures at the Collège de France, 1975-76*. Translated by David Macey. London: Penguin Books, 2004.

Foucault, Michel. *The History of Sexuality* Volume 1. Translated by Robert Hurley. New York: Vintage Books, 1990.

Gilbert, Pamela K. *Disease, Desire, and the Body in Victorian Women's Popular Novels*. Cambridge, Cambridge University Press, 1997.

Kete, Kathleen. "Introduction: Animals and Human Empire". In *A Cultural History of Animals in the Age of Empire*, edited by Kathleen Kete, 1–24. Oxford and New York: Berg, 2011.

Kreilkamp, Ivan. "Dying Like a Dog in *Great Expectations*". In *Victorian Animal Dreams: Representations of Animals in Victorian Literature and Culture*, edited by Deborah Denenholz Morse and Martin A. Danahay, 81–94. Aldershot: Ashgate, 2007.

Langland, Elizabeth. "Enclosure Acts: Framing Women's Bodies in Braddon's *Lady Audley's Secret*". In *Beyond Sensation: Mary Elizabeth Braddon in Context*, edited by Marlene Tromp, Pamela K. Gilbert, and Aeron Haynie, 3–16. Albany: State University of New York Press, 2000.

Mangham, Andrew. *Violent Women and Sensation Fiction: Crime, Medicine and Victorian Popular Culture*. Basingstoke and New York: Palgrave, 2007.

Mansel, Henry. "Sensation Novels". *Quarterly Review* 113 (1863): 481–515.

Miller, D. A. "Hitchcock's Hidden Pictures". *Critical Inquiry* 37, no 1 (Autumn 2010): 106–130.

Miller, D. A. *The Novel and the Police*. Berkeley and Los Angeles: University of California Press, 1988.

Miller, John MacNeill. "When Drama Went to the Dogs; or, Staging Otherness in the Animal Melodrama". *PMLA* 132, no 3 (May 2017): 526–42.

"Our Female Sensation Novelists". *Christian Remembrancer* 46 (1863): 209–36.

Prince, Gerald. "The Disnarrated". *Style* 22, no 1 (Spring 1988): 1–8.

Pykett, Lyn. *The Improper Feminine: The Women's Sensation Novel and the New Woman Writing*. London and New York: Routledge, 1992.

Review of *Armadale*. *Westminster Review* (October 1866): 269–71.

Ritvo, Harriet. *Noble Cows and Hybrid Zebras: Essays on Animals and History*. Charlottesville and London: University of Virginia Press, 2010.

Shukin, Nicole. *Animal Capital: Rendering Life in Biopolitical Times*. Minneapolis and London: University of Minnesota Press, 2009.

Siddiqi, Yumna. *Anxieties of Empire and the Fiction of Intrigue*. New York: Columbia University Press, 2008.

Steinlight, Emily. *Populating the Novel: Literary Form and the Politics of Surplus Life*. Ithaca and London: Cornell University Press, 2018.

Taylor, Jenny Bourne. *In the Secret Theatre of Home: Wilkie Collins, Sensation Narrative, and Nineteenth-Century Psychology.* London and New York: Routledge, 1988.

Warhol, Robyn R. "Neonarrative; or, How to Render the Unnarratable in Realist Fiction and Contemporary Film". In *A Companion to Narrative Theory*, edited by James Phelan and Peter J. Rabinowitz, 220–31. Oxford: Blackwell, 2005.

Wolfe, Cary. *Before the Law: Humans and Other Animals in a Biopolitical Frame.* Chicago and London: University of Chicago Press, 2013.

Woloch, Alex. *The One vs. the Many: Minor Characters and the Space of the Protagonist in the Novel.* Princeton: Princeton University Press, 2003.

“The Motto of the Mollusc”: Patricia Highsmith and the Semiotics of Snails

Sally West

Patricia Highsmith generally preferred animals to people. In her 1975 collection of short stories, *The Animal-Lover’s Book of Beastly Murder*, each tale narrates the successful revenge of an animal protagonist upon the cruel, ignorant and usually venal humans with whom they are forced to share their world. The collection includes the narratives of a cat, an elephant, some hamsters, but also those of creatures less conventionally likely to inspire sympathy, such as cockroaches and rats. Highsmith’s own choice of animal companions was similarly eclectic; as well as keeping a succession of cats, she had a particular affinity for snails. During her residence in Suffolk in 1966 she kept around 300 of the creatures as pets, sometimes taking them to cocktail parties and other events in an oversized handbag with a head of lettuce for them to feed on. When she moved to France the following year, she apparently smuggled some of her pets into the country under her breasts, making numerous trips in order to transport the whole menagerie.[1]

S. West (✉)
University of Chester, Chester, UK
e-mail: sally.west@chester.ac.uk

R. Hawthorn, J. Miller (eds.), *Animals in Detective Fiction*, Palgrave Studies in Animals and Literature,
https://doi.org/10.1007/978-3-031-09241-1_10

The fascination Highsmith had for her snails, and the apparent care with which she integrated them into her domestic and social life, has significance beyond being a quirky aspect of her biography. Perhaps unsurprisingly, snails also appear in two of Highsmith's short stories and a novel. In these fictional texts, Highsmith uses snails, and her characters' attitudes towards them, to expose the commodification of the individual; her snails, like her alienated protagonists, suffer the threat of being bought, sold, curated and consumed by capitalist culture. This aspect of Highsmith's work can be read as part of a continuum in detective fiction's treatment of animals as a vehicle for social and cultural criticism; in her representation of human/animal relationships, Highsmith is both conducting a searing critique of her specific culture and placing her work firmly in the traditions of detective fiction.

Patricia Highsmith's novels and short stories resist straightforward generic classification. This difficultly has been frequently recognised: her works "splay across genres";[2] she "never fitted comfortably into any generic category";[3] she "distorts and overturns every convention of crime fiction writing".[4] The novelist seemed actively to embrace her work's resistance to categorisation. While her "how to" writing guide may be titled *Plotting and Writing Suspense Fiction*, Highsmith takes issue there with the category in which her work has been most frequently pigeonholed in the United States, pointing out that "the suspense label does no one any good. In France, England and Germany, I am not categorized as a suspense novelist, but simply a novelist, with greater prestige, longer reviews and larger sales, proportionately, than in America".[5] Tzvetan Todorov's 1966 essay "The Typology of Detective Fiction" identifies the suspense novel as developing from the thriller, which in turn developed from the classic detective novel. It maintains the "mystery of the whodunit", but resists what Todorov calls a "simple detection of the truth".[6] Interestingly, however, Todorov specifically excludes Highsmith's most famous novel, *The Talented Mr Ripley*, from the sub-genre of suspense fiction, claiming that it is among a collection of novels "which have tried to do without both mystery and the milieu proper to the thriller … which are too few to be considered a separate genre" (52). In the decades since his essay, however, detective fiction has developed in ways which Todorov may have found an anathema to his argument that "the whodunit par excellence is not the one which transgresses the rules of the genre, but the one which conforms to them" (43). Highsmith's novels are now frequently taught on university modules in crime fiction, and the periodical *Clues: A Journal of Detection* published a special issue devoted to her work in 2015. This suggests that it is possible to identify ways of aligning Highsmith's writing with the tradition which started with the detective fiction of the mid to late-nineteenth century and which continues to proliferate and diversify in contemporary crime fiction.

It is true that the majority of Highsmith's novels include a crime of some description, but there are obvious semantic obstacles to calling them "detective fiction" in that relatively few include a conventional detective figure and, where they do, he is frequently a peripheral character, often having minimal involvement in the resolution of the story. Instead, Highsmith's focus is firmly on what Joan Schenkar calls "the ravaged psychologies, triturated egos, and fractured identities of her murderers" rather than "the dull mechanics of their deeds" (200). As Andrew Pepper points out, this "shift of perspective from investigator to criminal and from the 'social' to the 'psychological'" aligns Highsmith more with roman noir, the sub-genre of detective fiction which developed from shared characteristics of the American hard-boiled crime novel and film noir.[7] Some of the preoccupying themes of this sub-genre are, argues Pepper, "the corrosive effects of money, the meaninglessness and absurdity of existence, anxieties about masculinity and the bureaucratization of public life" (60). All of these features are observable in Highsmith's work, which frequently anatomises what lies behind the superficially civilised façade of middle-class, middle-American life. The "ravaged psychologies" of her murderous protagonists develop, somewhat paradoxically, as self-protective responses to a felt alienation in a culture which treats everything as a commodity to be bought and sold. One site for this conflict between the individual and their environment is in the treatment of, and attitudes towards the non-human animal, specifically, the snail.

In *The Poetics of Space*, Gaston Bachelard reflects on how the snail's shell is the ultimate evolving house, built by its inhabitant and growing and developing as the creature does. The "motto of the mollusc", he says, is "one must live to build one's house, and not build one's house to live in".[8] Like the snail, Bachelard argues, our relationship to the physical spaces we inhabit should be organic; our spaces should evolve as we do, reflecting our personality and psychology. Where this relationship is not reciprocal, alienation and stress result. Repeatedly in Highsmith's work, we encounter characters who, at least in part, define themselves through the spaces they inhabit; for them, as for the snail, to construct a personal space—a house, a room—is also to construct a self. However, this desired reciprocity between consciousness and space is frequently subjected to considerable stress from external forces, often representative of capitalist consumer culture. Highsmith's novels anatomise the uneasy relationship between self and world described by social anthropologist Daniel Miller: "people are constructed by their material world, but often they are not themselves the agents behind that material world through which they must live".[9] In Highsmith's novel *Deep Water*, snails function as an expression of the alienation and loss of self-determination experienced by her protagonist Vic Van Allen following repeated incursions into his physical—which is also his psychological—environment. Before turning to *Deep Water*, however, a consideration of two other representations of snails in Highsmith's short stories will help to establish some meanings of the mollusc. Dawne McCance observes that it is generally accepted in critical animal studies that "modern Western ways of knowing non-human animals ... have turned animals to 'stone', that is into inert objects, useful and disposable things".[10] In both stories, the human protagonists are driven by greed and narcissism; the snails are "objects" to be exploited and commodified for their use value. Highsmith, however, remaining firmly on the side of the gastropods, reverses the established human/animal power dynamic: her animal protagonists reveal the failings of the human ones, as they, in their turn, "use" and "dispose" of the men who would contain them.

Snail Watching: Highsmith and the Tradition of Animals in Detective Stories

"The Snail Watcher" narrates Peter Knoppert's fascination with snails. Initially, this new hobby seems to have a positive impact on him; he remarks "I never cared for nature before in my life ... but snails have opened my eyes to the beauty of the animal world".[11] His first experience of this "beauty", however, is initially rather perplexing; two of his snails were "behaving very oddly":

> Standing more or less on their tails, they were weaving before each other for all the world like a pair of snakes hypnotized by a flute player. A moment later, their faces came together in a kiss of voluptuous intensity. Mr Knoppert bent closer and studied them from all angles. Something else was happening: a protuberance like an ear was appearing on the right side of the head

> of both snails. His instinct told him that he was watching sexual activity of some sort (1–2).

Knoppert's instincts are correct about the nature of the activity he is observing. As the months go on, his collection of snails reproduces at an alarming rate, colonising his study. Knoppert spends his evenings watching them, feeding them and letting them crawl on his hands because "he fancied his snails enjoyed this human contact" (6). Knoppert's new hobby appears to have a beneficial effect on his work as well:

> His colleagues at the brokerage office noticed a new zest for life in Peter Knoppert. He became more daring in his moves, more brilliant in his calculations, became, in fact, a little vicious in his schemes, but he brought money in for his company. By unanimous vote, his basic salary was raised from forty to sixty thousand dollars per year. When anyone congratulated him on his achievements, Mr Knoppert gave all the credit to his snails and the beneficial relaxation he derived from watching them (5).

On the surface, it appears a little peculiar that the "beauty of the natural world" prompts an increase in Knoppert's involvement in the capitalist cycles which make money through potentially "vicious" dealings. This is where we perhaps need to reassess the nature of Knoppert's fascination. By this point, he has tired of watching the snails' intricate mating ritual, but still derives immense satisfaction from seeing the eggs hatch: "the thrill of seeing the white caviar become shells and start to move—that never diminished however often he witnessed it" (5). It is precisely the *proliferation* of snails which fascinates Knoppert—which has, in fact, become something of an addiction. Just as "*his* snails" (my emphasis) proliferate, so too does the money he makes for his company and the salary he claims as a consequence. Far from being softened and ameliorated by the beauty of nature, Knoppert plays out, in his pet gastropods, another form of capitalist accumulation: "He saw his bank account multiplying as easily and rapidly as his snails" (7). Knoppert's work at the brokerage is a part of the cycle of buying and selling, desiring and consuming, which is frequently anatomised in terms of its ill-effects in Highsmith's fiction. As such, it seems appropriate that the outcome she decrees for Knoppert in "The Snail Watcher" should also involve consumption—if of a slightly different sort. Unfortunately, Knoppert's commitment to financial proliferation causes him to neglect that of the gastropods. When he returns to the study after a two-week absence, the snails have taken over. Not content with colonising the room, however, they swiftly colonise Knoppert himself:

> A snail crawled into his mouth. He spat it out in disgust. Mr Knoppert tried to brush the snails off his arms. But for every hundred he dislodged, four hundred seemed to slide upon him and fasten to him again, as if they deliberately sought him out … There were snails crawling over his eyes … "Help!" He swallowed a snail. Choking, he widened his mouth for air and felt a snail crawl over his lips to his tongue. He was in hell! He could feel them gliding over his legs like a glutinous river, pinning his legs to the floor (8–9).

The last thing that Peter Knoppert sees is two snails copulating on a rubber plant and "tiny snails as pure as dewdrops … emerging from a pit like an infinite army into their widening world" (9). The study, previously Knoppert's private space, has become the "widening world" of the colonising army of molluscs and Knoppert, the facilitator of consumption, has now become the consumed.

Issues of colonisation and consumption are also apparent in Highsmith's other short story concerning snails, "The Quest for *Blank Claveringi*". Professor Clavering is an academic in his forties who "had achieved no particular renown" in his field of zoology, and as a consequence "very much wanted to discover some animal, bird, reptile, or even mollusc to which he could give his name".[12] Hearing some unsubstantiated reports of the existence of giant molluscs inhabiting a remote island, Clavering happily invests most of his capital in an expedition to verify their existence, and hopefully to bring one home. Rather like a more inept version of Walton, the protagonist of the framing narrative of Mary Shelley's *Frankenstein*, Clavering is a hopeless over-reacher, completely unprepared for what he discovers. His motives for the expedition, however, do not have even the pretentions of nobility or scientific endeavour of Walton, or, indeed, Frankenstein: he simply wants to claim a creature for posterity. The "*Blank*" of the story's title succinctly signifies that Clavering does not overly care what genus of creature he discovers, only that it will bear his name. This absence of discernment, the lack of real scholarly interest in the creatures he pursues, is significant, and we will encounter it again in *Deep Water*. In her fiction, Highsmith frequently identifies narcissism and carelessness as the deadliest of sins and punishes them accordingly.

The outcome for Professor Clavering, once he finds himself trapped on an island populated by two gigantic snails and their progeny, is a foregone conclusion. The hunter rapidly becomes the hunted and the story ends with an inevitability registered in Highsmith's description of the snails' movement towards the Professor: "there was something contemptuous, something absolutely assured, about the snails' slow progress towards him, as if they were thinking 'Escape us a hundred, a thousand times, we shall finally reach you and devour every trace of you'" (76). Finally, racing desperately towards the sea, Professor Clavering realises that "his fate was both to drown and to be chewed to death" (80).

These two gleefully macabre short stories offer some intriguing ways of reading the meaning of the mollusc in Highsmith's work. In both cases, the snails are resolutely alien, resisting human efforts to interpret and domesticate them. In "The Snail Watcher", Peter Knoppert has to resort to translating a note in French in Darwin's *Origin of Species*, referring to the work of naturalist Henri Fabre describing the mating ritual of snails, to establish with any certainty the significance of the "dance" which he observes in his study. Professor Clavering is mistaken in his belief that he is dealing with land snails and that the salt water of the sea will protect him from his pursuers: he is not, and it does not. In both stories, human knowledge of the creatures is tested and found wanting and the snails reveal the materialistic and narcissistic personalities of the human protagonists. For Knoppert, the satisfaction he takes in the proliferation of his snails reveals a careless lack of regard for unintended consequences: that the snails colonise his room and consume him is a reflection of his similar disregard for the possible effects of his "vicious" dealings in the world of finance. Growth—be it gastropodous or economic—is his sole concern. The question of size is also a feature of "The Quest for *Blank Claveringi*"; it is tempting to read the giant snails as representative of the size of the ego of the man who would hunt them down to generate a scientific footnote which would, in his eyes, confer a form of immortality upon him.

This conflict between human ego and untameable animal in these tales of hubris offers a way of situating Highsmith in the detective story tradition. If we consider some of the most well-known detective stories involving animals, Poe's "The Murders in the Rue Morgue" and Conan Doyle's *The Hound of the Baskervilles* and "The Adventure of the Speckled Band", some comparable features emerge. In all three, human actions, motivated by financial greed, transform an animal into a murder weapon. Driven by the desire to claim the Baskerville inheritance, Stapleton's treatment of the hound in Conan Doyle's novella breeds a ferocity and viciousness in the creature which is sufficient to kill two people. In "Speckled Band", Grimesby Roylott is, however implausibly, able to train the swamp adder to fatally poison one step-daughter, and is only thwarted in claiming the inheritance of the second due to the intervention of Holmes and Watson. In "Rue Morgue", although the sailor who brought the orangutan to Paris did not intend the deaths of the L'Espanayes, his aim to sell the creature for profit causes him to neglect its welfare and, by extension, that of the surrounding populace with whom it comes into contact. Furthermore, the creature's actions in the L'Espanayes' apartment are a result of behaviour learnt from the sailor and the fear engendered by his punitive use of a whip.

The hubris of Stapleton, Roylott and the sailor of "Rue Morgue" in believing that they can control and exploit animals for financial gain is developed in Highsmith's tales in the over-weening but ultimately misplaced confidence of her colonising and consuming human protagonists. In "The Snail Watcher", Knoppert, rather in the manner of a nineteenth-century collector, subordinates a free creature and transplants it into a foreign habitat; his snail aquarium is something of a neglected zoo. In this way, the snails' assault on his physical and bodily space could be read as a form of reverse colonisation. Similarly, Clavering is a coloniser of not only the island which the giant snails inhabit, but of the very fabric of their intimate home; it is after injuring one of the snails by piercing its shell that Clavering's fate is sealed. In this, both of Highsmith's protagonists recall the instincts of the sailor of "Rue Morgue", who transplants the orangutan from its native Borneo, and those of Roylott in "Speckled Band", who returns from the East with a menagerie of tropical creatures with which to populate the English estate of Stoke Moran.

Where Highsmith's presentation of snails differs from the function of the animals in these classic tales is in the autonomy that she grants the gastropods. In her stories, there is no ordering, resolving force in the form of a Holmes or a Dupin, and, rather than the threat, in the form weaponised animal, being contained at the conclusion of the stories, the beast remains centre stage, stubbornly resisting human control. In both stories, the protagonists' mistaken attitude towards snails reveals their failure to inhabit their worlds in a way that is careful or mutual. In "The Snail Watcher" and "The Quest for *Blank Claveringi*", the snails reveal the human desire to conquer and consume. Highsmith, who knew her Dante, constructs the appropriate *contrapasso* for her protagonists: each is conquered and consumed in their turn.

Living to Build One's House: The Value of Snails in *Deep Water*

Highsmith's merciless atomising of the faults of Knoppert and Clavering is reflected in a comment made by Graham Greene in his preface to her collection, *Eleven*, in which both of these stories appear: "Mr Knoppert has the same attitude to his snails as Miss Highsmith to human beings. He watches them with the same emotionless curiosity as Miss Highsmith watches the talented Mr Ripley".[13] If we turn to Highsmith's novels, Greene's comment can be qualified slightly: it is through the *perspectives* of the protagonists of her fiction—characters who often slide into psychosis and sometimes commit murder—that Highsmith is able to observe her mid-twentieth-century American society with the "emotionless curiosity" of the alienated, sometimes slightly bemused outsider. One example of this perspective, and another instance of a gastrophile, is Victor Van Allen, protagonist of her 1957 novel *Deep Water*. Vic's relationship with his snails is an embodiment of his wider rejection of the commodification and consumption he perceives both in his immediate social circle and in 1950s American society more widely. Following the motto of the mollusc, according to Bachelard, Vic lives to build his house.

On the face of it, Vic Van Allen is ordinary and a conformist. He has money, friends, a job he enjoys, a beautiful house and a beautiful wife. However, despite this superficial calm and comfort, many aspects of Vic's life can be read as very subtle rebellions against the capitalist culture and class structure that values the good job, the beautiful house and beautiful wife, but only as indicators of conformity, as indicators, in a materialistic sign system, of happiness. Victoria Hesford observes Highsmith's emphasis on the home as a site of dislocation and tension: "Highsmith's vision of 1950s America turns the Cold War domestic ideology of the middle-class home as a source of national strength and normality inside out, revealing the undertow of violence and sexual unconventionality that both prop up the public function of the middle-class home and constantly threaten to tear it apart".[14] It quickly becomes apparent in *Deep Water* that the "domestic ideology" which Vic presents to the world is, in fact, riven with conflict and contradiction. He is married to the beautiful and flamboyant

Melinda, whose own unhappiness takes the form of repeated, mostly disastrous affairs, which she flaunts in front of Vic and their young daughter Trixie. Vic has adopted a model of preternatural patience in the face of Melinda's infidelities, welcoming each of her lovers into the family home, supporting her when the liaison runs its course and generally presenting a calm, accommodating face to the world. Vic's endurance earns him some social capital in his community as a marital martyr, but the situation is more complex than this. Not only, as Peters observes, does Vic's "passivity cover a visceral distaste",[15] his apparent acquiescence in the face of Melinda's behaviour is merely one symptom of a coping strategy for what he perceives as a much wider cultural malaise.

Vic is aware that his friends are by turns impressed and frustrated by his tolerance of Melinda's behaviour, that they think him somewhat odd. But "Vic didn't mind at all being considered odd. In fact, he was proud of it in a country in which most people aimed at being exactly like everybody else".[16] He sees no point in being miserable about Melinda, because "[t]he world was too full of interesting things" (21), things which are, in fact, far more interesting than his wife. Highsmith implies that Melinda's wildness and non-conformity had been attractive to Vic in the past, but that these qualities have, since they married, been channelled into various forms of pointless sensual excess: throwing herself at unsuitable and frequently only barely interested men, drinking to excess only to kill her hangover with another succession of high-ball cocktails. There is a desperation about Melinda's behaviour which pains Vic, but in an aesthetic rather than an emotional sense. Frankly, she has begun to bore him: "She was not interested in anything he was interested in, and in a casual way, he was interested in a great many things—printing and bookbinding, bee culture, cheese making, carpentry, music and painting (good music and good painting), in star-gazing, for which he had a fine telescope, and in gardening" (20). To be interested by interesting things is, for Vic, a mark of culture and, crucially, a means of self-determination. For instance, his business is printing books, but not mass-market paperbacks. Vic's printing press prints only five or six books per year; unusual books by unusual people. Paper, type-face, cover designs are all original and pondered over at length. The business inevitably runs at a loss, but Vic has a private, inherited income which covers that. Essentially, the press is a hobby, but one which has earned Vic a reputation as someone who pays attention to intricate detail, someone who values unique and beautiful things ahead of vulgar concerns such as turning a profit. Vic's other interests include crafts of various forms, including carpentry:

> Vic took a cigarette from the box on the rosewood cocktail table. He had just made the table, had polished its very subtly concave top as carefully as if it had been a lens. He had made it to replace the old cocktail table which he had also made, that dated from Larry Osbourne and had become so stained with cigarette burns and alcohol, in spite of the protective waxes he had always kept on it, that he had had no desire to refinish it. He wondered how soon the rosewood table was going to be stained with rings from highball glasses and burns from neglected cigarettes (62).

The care with which Vic not only makes, but cares for this piece of furniture provides an insight into how he constructs his domestic space. Gaston Bachelard identifies the importance of "cherished" objects in a space:

> [W]hen a poet rubs a piece of furniture—even vicariously—when he puts a little fragrant wax on his table with the woollen cloth that lends warmth to everything that it touches, he creates a new object; he increases the object's human dignity; he registers this object officially as a member of the human household. ... Objects that are cherished in this way really are born of an intimate light, and they attain to a higher degree of reality than indifferent objects, or those that are defined by geometric reality. For they produce a new reality of being, and they take their place not only in an order but in a community of order (87–88).

Care of objects registers a relationship between an individual and the space that they share with such objects which increases the "human dignity" of both. Cherished objects "produce a new reality of being" according to Bachelard because they register our intimate selves *being* in the world. In peopling his home with unique, hand-crafted objects which are cared for like members of the family, Vic manifests himself in his physical environment.

Yet Vic must share this environment with those who do not respect the "human dignity" of such things: people who treat them as "indifferent objects", assessed only for functionality or material worth. The rosewood table encapsulates the estrangement between Vic and Melinda; it is an object which is imbued with the tension of their opposing sets of values. The old, spoilt table dates "from the time of Larry Osbourne", one of

Melinda's first lovers, indicating that the years of their marriage are marked out by the various manifestations of Melinda's infidelity and the physical damage of her carelessness with spilt drinks and neglected cigarettes. More than this: in Highsmith's emphasis on the intimate association between Vic and the tables he creates, we can read the damage done to one as damage done to the other. Like the table, Vic is also stained and scarred by the unhappy years of his marriage.

The table-making and nurturing is one example here of the care Vic takes of his physical environment, how he attempts to build it around him organically: in Bachelard's terms he "live[s] to build [his] house" (126). This is also manifested in the satisfaction he takes in domestic chores and regular maintenance and decorating so that, for instance, "[t]he living room looked as if happy people lived in it now, even if happy people didn't" (151). As well as peopling his home with objects which have human dignity, Vic also shares his domestic space with a collection of snails. Fiona Peters discusses the symbolic meaning of the snails in *Deep Water* in terms of the calm and order which they offer Vic in contrast to the brash, anxiety-inducing disorder of his marriage. Peters argues that "the emphasis on the world of the snails is not predicated on an alternative universe of meaning, but instead acts as a focus for the ways in which Vic, through his slide into psychosis, gradually subverts the symbolic (or meaningful) world that surrounds him".[17] Qualifying Peters slightly, it is possible to argue that while Vic's obsession with the gastropods can certainly be read as a marker of his increasing psychosis—he is, after all, a triple murderer by the end of the novel—the snails *do* in fact constitute "an alternative universe of meaning". If, as Philip Armstrong observes, the functions of animals in the modern novel have traditionally included "defining what it meant to be human as well as expressing ideas about sensibility and empathy",[18] then it follows that this "alternative universe" of snails is one with more moral and cultural currency than the prevailing culture which Highsmith depicts surrounding Vic in *Deep Water*.

Unsurprisingly, Vic and Melinda have separate rooms; in fact, Vic has a separate, if modest, wing, again self-constructed, adjoining the garage. In this space, Vic also houses and cares for his gastropods. Like Peter Knoppert in "The Snail Watcher", Vic has a fascination with the snails' mating ritual, yet unlike Knoppert their behaviour causes him a joy which does not diminish over time. His particular favourites are a pair he has named Edgar and Hortense who represent, for Vic, a sort of domestic ideal: "They mated about once every week, and they were genuinely in love, Vic thought, because Edgar had eyes for no other snail but Hortense and Hortense never responded to the attempt of another snail to kiss her ... That was true love, Vic thought, even if they were only gastropoda" (124). Here, the unavoidable interpretation of the snails is that the loyal, devoted Edgar is an idealised version of the tired and abraded Vic, and the virtuous Hortense is his fantasy, perhaps also his memory, of Melinda. However, there is a subtler, more complex reading to be had, of which Highsmith, as a committed gastrophile herself, could not have been unaware. The majority of snails are simultaneous hermaphrodites: they are male and female at the same time.[19] As such, Vic's idealisation of, and potential identification with Edgar and Hortense, presents suggestive ideas about sexual identity. Highsmith suggests that Vic has little sexual interest in Melinda; her affairs do not so much provoke sexual jealousy as domestic irritation. As Hesford points out, Vic's feeling of increased vigour after murdering one of Melinda's lovers manifests itself not in sexual desire but in an urge to vacuum the house (224–225). We have seen how he orders and cherishes his domestic space, and how the visible presence of disorder within it is painful to him. It is possible, then, that part of Vic's fascination with his snails derives from a fantasy of domestic sexual equity which goes beyond prescribed gender roles and the strictures of conventional masculinity. The snail, which grows its own home organically, and which can adopt either biological sex, thus becomes a paragon of domestic perfection. It is notable that Vic's admiration for the mutual devotion he detects in the relationship between Edgar and Hortense intensifies as the novel develops, in inverse proportion to the increasingly unavoidable reality of the disaster that is his marriage.

As with the rosewood table, however, Vic's reading of the snails must compete with those of others, in this case, the reading of Tony Cameron, Melinda's final lover and Vic's second victim. Following the usual pattern, Melinda frequently brings Cameron back to the house for meals, drinks and music sessions late into the night. Throughout, Vic endures this incursion into his territory and attempts to perform the role of genial host. The boorish Cameron invades not only what might be regarded as the communal areas of the house—the lounge, the kitchen—but, on one occasion, Vic's private space, the garage where he keeps his snails. Vic's careful and delicate cleaning of the snail aquaria sits in stark contrast to Cameron's crude interference:

> Delicately, he [Vic] detached a young snail that had crawled up on the part of the glass he was washing, and set it down on the earth inside the tank.
>
> "Look like they'd be good to eat", Mr Cameron remarked.
>
> …
>
> He [Vic] looked over at Mr Cameron and said "I wish you wouldn't take the screen off, if you don't mind. They crawl out very easily".
>
> Mr Cameron straightened up and slid the screen top back with a carelessness that made Vic wince, because he felt sure that a baby snail or two must have been crushed. Mr Cameron probably hadn't even seen the tiny baby snails. His eyes didn't focus that small (205).

Highsmith deliberately contrasts the care with which Vic handles the young snail to Cameron's "carelessness" towards the gastropods. What is at stake here is a question of discernment: while Vic finds beauty in detail, in the miniscule, Cameron's "eyes didn't focus that small". Moreover, to discern is not only to see clearly, but to discriminate. All of Vic's activities, from his snail-rearing, to his work at the printing press, to his completion of domestic tasks are marked by discrimination and attention to detail. Cameron, on the other hand, despite liking music, plays the clarinet with "no quality … he emanated noise" (208). A lack of discrimination is a characteristic shared by all of Melinda's lovers and by implication, Melinda herself. As Vic reflects, "It was not that he objected to Melinda's having affairs with other men *per se*, … it was that she picked such idiotic, spineless characters" (18).

The morally spineless lover and physically spineless gastropods are brought into opposition once again when Cameron suggests that they eat some of Vic's snails:

> "The snails are not for eating", Vic said.
>
> Cameron's face fell a little. "Oh. Well—what the hell are they for?" he asked, laughing. "Melinda said—"
>
> "I don't use them for anything. They are useless", Vic said, spitting the words out with a particular bitterness (228).

From Cameron's perspective, the snails' potential to be converted from living creature to edible foodstuff renders them a commodity to be consumed. In contrast, Vic rejects the perspective that the snails have a use-value: *he* does not *use* them; they are "useless". This tension in viewpoints places Highsmith's representation of the snails at the heart of an argument about capitalist culture. Vic prides himself on his nonconformity in this respect: "his nonprofitable publishing business, ... his television-less household, and perhaps even his superannuated car" (278). He styles himself as more original, more cultured, unsullied by the grubby taint of capitalist profit.

In a discussion of *Disgrace*, Philip Armstrong suggests that Coetzee's novel prompts the question "How should we respond to those, humans and nonhumans, who have been left dispossessed and 'surplus to requirements' in the aftermath of colonialism and industrial modernity?" (46). *Deep Water*, too, meditates on the culturally "surplus" and "dispossessed" who resist the colonisation of consumerist modernity. Vic cannot win the battle of the non-conformist in a world of conformity; he cannot, finally, live to build his house, much less his world. His tolerance of Melinda's affairs turns out to have a limit, but what he cannot tolerate is less the knowledge that his wife is sexually unfaithful than the intrusion of a succession of crass, undiscerning individuals into his ordered home. If, as has been suggested, we can detect a relationship between Vic's domestic spaces and his psychology, then these physical incursions are also psychological invasions, each bringing another virulent influx of careless consumption, each chipping away a little more at his sanity. In the exchange about the use-value of snails, Vic's disgust with Cameron and his ilk is crystallised. They—the conformers—value "use" over beauty, profit over job satisfaction, the latest television over a hand-crafted table and sensual indulgence over cultural pursuits. One such conformer is Wilson, the acquaintance of the Van Allens who is instrumental in Vic's final arrest. By the end of the novel, Vic has committed three murders in futile attempts to defend his world from the conformers. After his final murder, that of his wife, he is led away by the police from the home which was, but is no longer, a manifestation of himself:

> [F]eeling very calm and happy, Vic kept looking at Wilson's wagging jaw and thinking of the multitude of people like him on the earth ... and thinking that it was not bad at all to be leaving them. The ugly birds without wings. The mediocre who perpetuated mediocrity, who really fought and died for it. He smiled at Wilson's grim, resentful, the-world-owes-me-a--living face, which was the reflection of the small, dull mind behind it, and Vic cursed it and all it stood for. Silently, and with a smile, and with all that was left of him, he cursed it (296).

It is one of Highsmith's great skills as a novelist that she can render a protagonist's defeat as a form of victory. Vic is the underdog, the non-conformist, the man who would stand up against "the mediocre who perpetuated mediocrity". Yet he has little left, and uses what he has in a curse upon the "ugly birds without wings". It is perhaps significant that at the denouement of this novel Highsmith chooses an animal metaphor to register the disjunction between her protagonist and the community from which, in the tradition of the closing moments of classic detective fiction, he has been ejected. With the protective shell of his home smashed open, Vic takes to the skies in the implicit contrast he draws between himself and what remains of his community. The physical incarceration which doubtless awaits him cannot, he suggests, contain the psychological agency of the bird with wings.

In *Deep Water*, Highsmith's use of the snail functions as a means of articulating dislocation, alienation and anti-capitalist sentiment. The motif of the snail's shell as an organic, evolving home, as expressed by Bachelard, offers an alternative to the fabricated, grafted-on trappings of consumer culture, but it is not, finally, an alternative which can be sustained. The mainspring of detective fiction generates from the severing of the social contract which dictates that the community functions as long as the individual abides by its rules. Highsmith is thus working within this tradition, in that she focuses on the conflict which arises when an individual psychology is pitted against a communal one; she recognises that such a conflict can only have one outcome, and Vic is apprehended by the communal forces of law and order. Where she differs from tradition, however, is in her rendering of the relative merits of the individual perspective versus that of the community. As she wrote in one of her notebooks, or "cahiers", "The psychopath of a book is an average man living more clearly than the world permits him".[20] Concluding *Deep Water* as she does, with Vic Van Allen's arrest, but also with his vigorous articulation of the value of the non-conformist perspective, Highsmith both adheres to and inverts the trope of detective fiction which Auden termed the "fantasy of being restored to the Garden of Eden".[21] In doing so, she reveals the delusion that lies at the core of the hope that the removal of one bad apple will heal the wounds of an entire community, and "expos[es] the constraints of normality for the hypocritical straightjacket she believed it to be".[22]

Coda: The Mollusc Lives On

In 1969, Highsmith was thinking about another story concerning snails. It is unfortunate that she did not, in the end, write the piece, as the synopsis of it that exists offers intriguing possibilities, both for complicating further the range of genres in which she worked and for offering a rather different perspective on snails and space. This story would have been about snails *in* space. Highsmith's biographer, Andrew Wilson, describes the scenario:

> [The story would be] focusing on an apocalyptic, post-nuclear world in which all life on the planet had been destroyed except for snails. A spaceship carrying the last 150 members of the human race arrived on the planet and they set out to destroy the gastropods, many of which have mutated—some creatures sport two heads or have grown to giant proportions, some possess remarkable intelligence, while others have developed cannibalistic habits. The battle between the snails and the humans is a fierce one—the creatures proceed to attack the men and eat them—but a few people manage to flee back into the spaceship and escape. Unknown to them, however, there is a small batch of snail eggs on board (268).

Armstrong asserts that "animals in texts tend to escape from our symbolic readings of them" as such representations always express "the kinds of practices, behaviours and interests that shape our actual, material interactions with them" (43). In Highsmith's projected post-apocalyptic story, her snails escape containment both within the plot and in terms of any critical efforts to read them purely symbolically. While issues of colonisation, consumption and the question of making a space in the world continue here, the balance of power has shifted and her snails maintain a glorious animal otherness. Just as the animals which appear in the early detective fiction of Poe and Conan Doyle escape, literally and symbolically, from the control of the humans who would exploit them, in her fiction involving snails, Highsmith demonstrates that ideologically, morally and sometimes even physically, the non-human animal triumphs over the human who would contain it.

Notes

1. Wilson, *Beautiful Shadow*, 267.
2. Schenkar, "Patricia Highsmith", 203.
3. Peters, "Introduction: Re-evaluating Patricia Highsmith", 5.
4. Bradford, *Crime Fiction*, 41.
5. Highsmith, *Plotting*, 141.
6. Todorov, "The Typology of Detective Fiction", 50.
7. Pepper, "The American roman noir", 59.
8. Bachelard, *The Poetics of Space*, 126.
9. Miller, *Stuff*, 84.
10. McCance, *Critical Animal Studies*, 2.
11. Highsmith, "Snail Watcher", 1.
12. Highsmith, "*Claveringi*", 63.
13. Greene, Foreword to *Eleven*, xi.
14. Hesford, "Patriotic Perversions", 217.
15. Peters, *Anxiety and Evil*, 48.
16. Highsmith, *Deep Water*, 19.
17. Peters, *Anxiety and Evil*, 55.
18. Armstrong, "Animating the text", 45.
19. Janssen and Baur, "Seasonal effects", 2917.
20. Highsmith, quoted in Wilson, *Beautiful Shadow*, 196.
21. Auden, "The Guilty Vicarage", 4.
22. Peters, *Anxiety and Evil*, 151.

Works Cited

Armstrong, Philip. "Animating the text: human-animal relations." *English in Aotearoa* 65, (July 2008): 41–48.

Auden, W. H. "The Guilty Vicarage." *Harper's Magazine*, (May 1948). http//harpers.org/archive/1948/05/the-guilty-vicarage/.

Bachelard, Gaston. *The Poetics of Space*. Translated by Maria Jolas. New York: Penguin, 2014.

Bradford, Richard. *Crime Fiction: A Very Short Introduction*. Oxford: Oxford University Press, 2015.

Greene, Graham. Foreword. *Eleven* by Patricia Highsmith. London: Bloomsbury, 2007.

Hesford, Victoria. "Patriotic Perversions: Patricia Highsmith's queer vision of Cold War America in *The Price of Salt*, *The Blunderer*, and *Deep Water*." *Women's Studies Quarterly* 33, no. 3 & 4 (Fall/ Winter 2005): 215–233.

Highsmith, Patricia. *Deep Water*. London: Virago, 1993a.

Highsmith, Patricia. *Plotting and Writing Suspense Fiction*. London: Sphere, 1993b.

Highsmith, Patricia. "The Quest for Blank *Claveringi*." In *Eleven*. London: Bloomsbury, 2007a.

Highsmith, Patricia. "The Snail Watcher." In *Eleven*. London: Bloomsbury, 2007b.

Janssen, Ruben and Bruno Baur. "Seasonal effects on egg production and level of paternity in a natural population of simultaneous hermaphrodite snail." *Ecology and Evolution* 14, no 5, (2015): 2916–2928.

McCance, Dawne. *Critical Animal Studies: An Introduction*. New York: State University of New York Press, 2013.

Miller, Daniel. *Stuff*. Cambridge: Polity, 2009.

Pepper, Andrew. "The American roman noir." In *The Cambridge Companion to American Crime Fiction*. Edited by Catherine Ross Nickerson. Cambridge: Cambridge University Press, 2010.

Peters, Fiona. *Anxiety and Evil in the Writings of Patricia Highsmith*. Farnham: Ashgate, 2011.

Peters, Fiona. "Introduction: Re-evaluating Patricia Highsmith." *Clues: A Journal of Detection* 33, no 2, (2015): 5–8.

Schenkar, Joan. "Patricia Highsmith." In *The Cambridge Companion to American Novelists*. Edited by Timothy Parrish. Cambridge: Cambridge University Press, 2012.

Todorov, Tzvetan. "The Typology of Detective Fiction." In *The Poetics of Prose*. Translated by Richard Howard. Oxford: Blackwell, 1977.

Wilson, Andrew. *Beautiful Shadow: A Life of Patricia Highsmith*. London: Bloomsbury, 2010.

"Before the white man came, when animals still talked": Colonial Creatures in Sherman Alexie's *Indian Killer* and Adrian C. Louis's *Skins*

Alexandra Hauke

Introduction: We Are All Animals

In his seminal essay "A Seat at the Table: Political Representation for Animals", Cherokee scholar Brian K. Hudson contends that animals "not only hold significance culturally and ethically, but also continue to occupy a place of importance in Native ways of understanding politics".[1] Hudson maps out a space for the relationship between humans and animals in Indigenous epistemologies in which no organism is superior but rather part of a synergy between creatures who share an interest in ecological balance, sovereignty, and justice. He goes on to emphasize that "many Native traditional stories teach us that we should realize with humility our place as one of many species" and thus as part of a more comprehensive

A. Hauke (✉)
University of Passau, Passau, Germany
e-mail: Alexandra.Hauke@uni-passau.de

R. Hawthorn, J. Miller (eds.), *Animals in Detective Fiction*, Palgrave Studies in Animals and Literature,
https://doi.org/10.1007/978-3-031-09241-1_11

ecosystem, wherein animals can function as much as politically motivated agents as human beings.[2] Such narratives, among them tribal creation stories, challenge what Wendy Woodward and Susan McHugh term the "classifying imagination of European settler colonialism";[3] they play a major part in tracing the natural, circular, and fluid balance of Indigenous identities as well as the cultural, philosophical, and religious meanings and values of animal beings in Native American world views. From this it follows that the foundational function of animals is invoked through their "crucial roles as creators in Native American sacred stories".[4] In such narratives, as Patricia Ann Lynch and Jeremy Roberts explain, "animals act as messengers, guardians, advisers, and servants of humanity"; some accounts even "say that before humans came into the world, animal spirits ruled over the Earth and animals could speak. When animals turned the Earth over to people, they lost their ability to speak. Many people, however, could understand animal languages and communicate with animals".[5] The educational value of this kind of storytelling thus lies in an awareness of animals as more than hunting trophies or inferior objects. A regard for their cultural and ethical significance can thus cut across Western deliberations about "how we are different from animals". It can also decentre the associated divide between humans and animals, on the one hand, and the imagined linear trajectory of lives that include birth and death as absolute beginning and end points without opportunity to move on to metaphysical realms and states of being, on the other hand.[6]

In this chapter, I want to carve out the narrative and political processes of transgression between human and animal forms in stories that pertain to the understudied genre of Native American detective fiction.[7] Sherman Alexie's *Indian Killer* (1995) and Adrian C. Louis's *Skins* (1995) do not eschew the problematics of binary understandings of identities or species which continuously cast Indigenous peoples and animals as part of an inferior order because they do not conform to the expected understandings of cultural and cognitive hierarchies set in place by and for Western philosophies of humanity. Instead, the texts highlight how the fluidity of existence of all beings leads to a more circular understanding of physical and metaphysical worlds across all temporal and spatial planes. They do so by engaging in discussions of the significance of animal spirits for human agents, and vice versa, in Indigenous creation stories. Through their focus on law enforcement and the limitations of political freedom for many Native individuals, both novels also approximate generic, thematic, and political strategies of detective fiction.

Indian Killer and *Skins* reimagine the cultural-political functions of animals in the Indigenous detective genre, which emerges as a transgressive literary form that simultaneously exposes and makes use of Western conventions of detective writing to open discussion regarding the legal processes that both frame and often juxtapose federal and Indian law. Native American detective fiction thus plays with the constraints of literary form, not least because of its narrative focus on the political and legal dimensions of human–animal relationships at the core of Indigenous epistemologies and thus beyond linear murder-suspect-solution plot lines. I discuss how Alexie's and Louis's novels approximate processes of rewriting genre fiction before emerging as newly envisioned versions of a well-known literary form from the perspective of the colonized other. Thereby, they debunk myths such as those about "the white man's Indian"[8] in the narratives of American mainstream culture, in which Natives are often imagined as uncivilized, beast-like savages subordinate to their human masters, or idolized for their free-spirited essence and innate connections to nature.[9]

Indian Killer and *Skins* have both been praised for their honest depictions of their respective protagonists' struggles for survival on and off their native reservations, for their negotiations of the hardships of poverty and discrimination, and their efforts to challenge the dominant—and wrongful—images of Indigenous peoples as presented in American Westerns and other genre fictions. James Giles argues that, with *Indian Killer*, Alexie is "deconstructing the detective novel form" because "the identity of the Indian Killer is never revealed";[10] the novel thus comes close to, on the one hand, what Merivale and Sweeney term metaphysical or anti-detective fiction in that it escapes conventional closure and does not rely on any single detective figure as a central investigative authority. Jeff Berglund, on the other hand, conceives of the novel as a "postmodernist serial killer thriller" whose selling point lies in scrutinizing the reader's assumptions about the double meaning of the title: "killer who is Indian or killer of Indians".[11] *Skins*, by contrast, has so far escaped theorizations of detective fiction or its sub-genres, arguably because it does not so clearly enter into conversation with the plot structures and devices of the detective tradition; rather, it participates in discussions of Lakota spirituality, trickster discourse, and resistant practices to colonial oppression. In this sense, both texts, like other Indigenous detective novels, expose normative understandings of literary form and thereby scrutinize the practices of generic categorizations.[12] This manner of circumventing yet simultaneously investing in certain signs of detective forms defines my understanding of *Indian Killer* and *Skins* as unambiguous examples of Native American detective fiction, where the marker "detective" does not function as an exclusive referent to any conventional investigator, because such a figure is absent in both tales.

Rather, Alexie's and Louis's novels *are* detective; in an adjectival sense of the term, the texts' efforts to engage in the processes of detection and detecting function in ways independent of most formal-aesthetic, textual, or narrative techniques of the detective genre.[13] The central pursuits of *Indian Killer* and *Skins* are thus neither the protagonist's triumphant capture of a criminal nor the restoration of the social status quo; instead, the narratives investigate occurrences and impulses of crime and violence at the hands of US imperial forces and the protagonists' opportunities and limitations as beings caught between physical and metaphysical planes as well as between Native and Western realities. Both main characters are also confronted with the transformative power of animal mythologies, spiritualities, and theologies, not least through the significant roles tricksters and other sacred animals play in their journeys of survival and self-discovery. Dave Aftandilian has observed that animals in many tribal creation stories "epitomize the different cosmological realms of existence, and some of them even have the ability to cross between these realms"[14]—practices that also shape the human–animal interactions in *Indian Killer* and *Skins*. In this way, the respective protagonists become injected with animal energies while remaining in their human forms, mediating between the dimensions without the need for anthropomorphism while, like many Native peoples, "deeply dependent upon the animals for both physical sustenance and spiritual knowledge and power".[15]

Ultimately, both novels dissect the textual possibilities of detective fiction and the idea of human-above-animal superiority: they emphasize certain generic features only to subvert them in the same breath; at the same time, the protagonists might be humans, animals, or both at different times or in different dimensions. No matter their form, however, they always remain racialized victims of the hegemonic beliefs of the US nation state and, thereby, objects of violence and racism: the former through their ambiguous designation as the Indian Killer and lack of knowledge about their cultural heritage; the latter through the abysmal yet inescapable living conditions on the Pine Ridge Reservation. As Cary Wolfe suggests, "you can't talk about race without talking about species, simply because both categories—as history well shows—are so notoriously pliable and unstable, constantly bleeding into and out of each other. Exhibit A here, of course, is the analogy between humans and animals".[16] In *Indian Killer* and *Skins*, the transgression of species and textual boundaries serves as an attempt to counter and break apart the racial and generic epistemes. At this point, both novels display their political power: the spiritual energies the protagonists summon from the animal world are initially employed to avenge the injustices experienced by Indigenous peoples as domestic-dependent nations and to re-narrate tribal history from the point of view of its makers, who, in many Indigenous creation stories, are animals. As such, while *Indian Killer* ends in a "Creation Story" and thus triggers seemingly infinite (re)births of the eponymous killer, *Skins* closes with the main character's vandalism of Mount Rushmore, an instance of overwriting the colonial imposition of white history.

In both novels, acts of vengeance by the colonized on the colonizer become acts of creation through the efforts of human-animal agents who are part of a natural ecosystem in which different species—and ethnicities—are encouraged to coexist. Creation thus transcends detection. As such, Alexie and Louis, rather than deconstructing Western conventions of detective writing, create newly envisioned versions of a genre in which the employment of animals as conscious, influential agents reminds the reader of a time "before the white man came, when animals still talked"[17]—a time when tribal communities lived uninhibitedly according to their creation stories and celebrated their hybridity as beings between the physical and spiritual planes, between animal and human forms. Creation "is not just an event that happened in the past, as described in Genesis, but is an ongoing and distinguishably human-animal process"[18] that cuts across the Western course of linearity and the singular idea of existence as either human or animal but never both or neither. Gerald Vizenor argues that such fluidity "creates a *presence* of animals and nature" in Indigenous literatures "with tropes and descriptions that are not bound to the modes of scientific causation or objective representation".[19] The novels under consideration ensure the continuance of Native peoples and their stories through such animal presences and through reconsiderations of literary form and Indigenous histories, in which human–animal relationships define the cycles of nature. Vizenor further stresses that the animals in Indigenous fictions "are connected to the environment, not the similes of human consciousness"; these creatures, which he terms "authored animals", display a "*natural* character" and counter one-dimensional figurations of anthropomorphized animals who simply "assay human characteristics".[20] The strength of authored animals lies in their motion and endlessness, "their presence and motivation".[21] In what follows, I argue that authored animals in Native American detective writing break down binary concepts of the Western imagination to foreground the circular ethos of tribal epistemologies, wherein creation is forever ongoing. The boundaries of existence and literary form thus become porous, melting into each other to create multidimensional mosaics of relationships beyond colonial lines.

"The tree grows heavy with owls": Animal Mythology in Sherman Alexie's *Indian Killer*

Sherman Alexie's 1996 novel *Indian Killer* has been read from various generic vantage points: Meredith James sees it as a "reverse captivity narrative";[22] Michelle Burnham reads it "as an instance of Indigenous gothic";[23] James Giles initially calls it a "parodic variation of the detective novel" and ultimately classifies it as an "anti-detective novel";[24] and Ray B. Browne includes it in his study of American Indian crime fiction, yet claims that it is "not a detective mystery".[25] These diverging classifications testify, on the one hand, to the inherently hybrid nature of genre both within and beyond the context of detective fiction. On the other hand, they draw out the common practice of reading literary works by writers of marginalized or racially marked groups against established Western categories and thus defining them by what they are decidedly not. While *Indian Killer* arguably shows traces of the aforementioned categories, I want to focus on the ways in which it affirms notions of a specifically Indigenous tradition of detective writing that does not forcefully participate in textual opposition but rather in political resistance.

Throughout the narrative, the eponymous murderer, the so-called Indian Killer, terrorizes Seattle while the ironically named Indigenous adoptee John Smith struggles with questions of belonging. Both the killer and John emerge as cultural hybrids and as individuals who continuously dis- and reappear across the scenes of the novel without being given the possibility to settle down. Alexie never reveals who the murderer is or if they are in fact Indigenous; questions like "what makes you think the Indian Killer is Indian anyway?" or "If it was an Indian doing the killing, then wouldn't he be called the Killer Indian?" (247) are never answered. In this context, the assumption about the killer's male gender is in part attributed to Alexie's skill as "a good detective writer", who "plants clues throughout the novel that seem to identify John as the killer",[26] merging their initially separate plot strands and exposing the dominant imagination of the racialized perpetrator. The killer roams the streets of Seattle seemingly unimpeded, their only care a desire to kidnap and murder white men in killing sprees inspired by a thirst for revenge against the colonial oppression of Indigenous tribes. Both John Smith and the Indian Killer believe that all Indigenous communities can be avenged through their respective efforts, a misjudgment represented in the circumstance that fighting crime with crime ultimately strips the novel of any chance for resolution. The fact that the killer is never found and slips through the authorities' fingers several times is explained through their connection with an owl, a bird of prey, whose silent hunting and skilled killing practices, ability to fly quickly from the scene, and camouflaging plumage the criminal adopts and translates into human form in gruesome ways. During an investigation at one of the novel's crime scenes, an officer in charge is reading a novel instead of paying attention to their deeds; they "never looked up as the killer passed within two feet of him" (299). Consequently, and after "[s]ilently singing an invisibility song learned from a dream, the killer carried the body [of their first victim] to an empty house" to perform what would become their signature ritual:

> Kneeling beside the body, the killer cut the white man's scalp away and stuffed the bloody souvenir into a pocket. (...) With hands curved into talons, the killer tore the white man's eyes from his face and swallowed them whole. The killer then pulled two white owl feathers out of another pocket, and set them on the white man's chest (54).

The killer's equivalence with the owl transforms them into an animal-predator who claims to seek justice for their people in the form of "private (...) revenge" that quickly turns into "an epidemic of anger"[27]—the principal affect at the core of the narrative.[28]

The killer's admiration for the owl's lack of remorse after each kill and obsession with its symbolism as a messenger of death leads them to misappropriate the animal's innate instinct to kill for food for their own purposes: "It lived to hunt, and hunted to live" (149), the killer muses. "The owl hunted to eat. It had no message" (192). The killer does not hunt to live; they kill because they "wanted people to know about the message of the knife, and knew who would be the messenger" (192). The process of choosing their victims becomes more and more deliberate in the context of this imagined message to the world, an arguably psychotic act that uncovers a link between the killer's exploitation of both the owl and their targets as symbols. Hudson argues that much of tribal storytelling "recognizes the suffering of animals being hunted for sport or profit";[29] the Indian Killer turns this fact on its head by hunting certain humans to make them suffer in retaliation for their crimes on animals and Indigenous peoples. The killer's acts of vengeance thus become symbolic in a different manner: their assumption of the owl's alleged habits, that it "wasted neither time nor emotion" (149) and that it "kept no souvenirs, no mementos from the scene of the crime" (192), suggests the killer's desire for rationality in the event of murder, a carelessness they equate with the ruthlessness of the all-powerful white man to blame for all colonial crimes on Native peoples. In adopting such behaviours, the killer imagines a self-induced liberation from the pain of their past and present and, thereby, a future free from being hunted, imprisoned, racialized, and discriminated against. However, like the novel as a whole, the killer's calculated manoeuvres promise no (re)solutions: their killing sprees neither absolve them from their rage nor provide their life with purpose; they do not "change the world" (192) in the way they envision. Like John Smith, whose hopeless efforts to discover himself and his heritage end in suicide, the killer instead remains "trapped in an urban wilderness",[30] whereby they become an allegory of a bird, a creature of the designated natural world, imprisoned in a cage, in turn a construct of the culture that defines Seattle.

The historical binaries of the Western imagination between animal and human, between savage nature and civilized culture, and, by extension, between the "Indian other" and the white man are in full effect at this point in the novel. The cosmopolitan backdrop and urban cityscape of Seattle hence becomes a prime symbol of these dualisms: overrun by crime and violence, and defined by colonialism and decay, the metropolis strips Indigenous peoples of their significant ties with each other, their traditions and their rights, in favour of the construction of "the last skyscraper in Seattle" (24), a building project executed by the city's biggest construction company, which also serves as John Smith's employer. The "last" skyscraper thus serves as an image of impending doom that characterizes both the suggested downfall of the city due to its high crime rates and construction overhaul as well as the connection between John and the killer. Both find themselves at a point of no return: while the killer feels "incomplete" (149) despite their bloody deeds, John is equally lost due to his missing family links and just "wanted to be a real person" (19). Both intend to solve their uncomfortable states of being in the same way, namely by murdering white men who they deem accountable for their individual and their felt collective Native pain.

In the context of Seattle's urban violence against its Indigenous population, the killer's self-identification with the owl signifies a deconstruction of the enforced binary between human and animal through violent means: when they kidnap six-year-old Mark Jones, the reader is confronted with images of the killer as Big Owl Man, "one of the central Native American 'bogeymen' that is used to instill fear, especially in children", either in "humanoid (human-like) form" or as "an owl-shaped monster who hunts children and other victims as prey".[31] In the same way that Big Owl Man is a "shape-shifter because it can change form between a man and an owl", the killer continuously uses their human–animal traits and skills to their advantage; Mark Jones even describes his kidnapper as "a bird. [...] I think it could fly because it had wings" (324). At once terrified of the kidnapper as Big Owl Man and fascinated by their ability to assume the role of an animal, the boy points toward a relationship between both the killer and John Smith with the bird of prey. In one of the last chapters of the novel, entitled "Flying", John, at this point dangerously close to echoing the killer's thirst for revenge against white men, steps off the so-called last skyscraper in Seattle and falls forty floors, flies through the air, and observes himself in an out-of-body experience after hitting the pavement before walking away, seemingly continuing his life after death on a plane removed from but still connected to the physical dimension. The novel's final chapter, entitled "A Creation Story", emphasizes this fluidity between the realms, all the while bringing John's unfathomable heritage and the killer's undissolved identity full circle. The Indian Killer participates in a sacred dance ceremony on a "cemetery on an Indian reservation. On this reservation or that reservation. Any reservation, a particular reservation" (419), confronting readers with the flippant Western stereotype that all Indigenous tribes or peoples are equal and indistinguishable. While this ceremony invokes the sacred Ghost Dance of Wovoka, a nineteenth-century religious movement, through which Native dancers could reunite the living with the dead in a metaphysical realm that will forever be part of the cosmological ecosystem, in *Indian Killer* it is imagined as an "owl dance", tracing the arc to the novel's animal symbolism. In an earlier scene, John cannot understand how and why tribal peoples who fear the owl as a messenger of death

> still owl danced. [...] Were the Indians dancing out of spite? Were they challenging the owl? With Indians, death was always so close anyway. When Indians owl danced, their shadows were shaped like owls. What was one more owl in a room full of Indians dancing like owls? (36)

The owl dance John observes "expresses the nearness of death, the tragedy haunting the tribal group that dances, but it simultaneously asserts the continuity of life".[32] In this sense, when John ends his life and thereby his time on the physical plane, his owl dance—a metaphor of the ceremonial Ghost Dance—allows him to pass on, to "let the wind take" him away, to "float away" (412), possibly to a realm where he can finally experience the life he dreamed of during a chapter entitled "How He Imagines His Life on the Reservation" (48). At this point readers are yet again reminded of the novel's last chapter, during which the killer is ultimately recreated as a new iteration of Wovoka in a powerful ceremony meant "to drive the white man into the sea":[33]

> The killer is softly singing a new song that sounds exactly like an old one. As the killer sings, an owl silently lands on a tree branch nearby. The owl shakes its feathers clean. It listens. The killer continues to sing, and another owl perches beside the first. [...] The killer sings and dances for hours, days. Other Indians arrive and quickly learn the song. A dozen Indians, then hundreds, and more, all learning the same song, the exact dance. [...] With this mask, with this mystery, the killer can dance forever. [...] The tree grows heavy with owls (419–20).

In the end, the killer is never found and never (not) identified as John Smith; they displace any efforts on behalf of the novel's investigators and thus ultimately frustrate the text's final resolution. The killer is imagined as an immortal embodiment of the owl as "death itself" (37), signifying a tribal "concept of an afterlife" in which "animal souls [...] travel to the Darkening Land (*Usunhi-yi*) after death" to live on.[34]

Within the frame of *Indian Killer* as a detective text, an open ending where the killer "plans on dancing forever" and consequently "never falls" (420) suggests that the United States continually deters Indigenous peoples from finding sovereignty and justice. Alexie's narrative thus imagines not only the killer and John Smith but also animals such as the owl as what Woodward and McHugh call "fitting symbols and extensions of lives under siege", at the mercy of the US justice and political systems. [35] Both

John Smith and the Indian Killer are initially visualized as birds with clipped wings, as colonial creatures robbed of their identities, whose engagements in revenge ceremonies ultimately enable them to fly away from the physical realm and reemerge through (re)creation on another spiritual plane. Which form they might take we will never know, a fact that serves as a final metacommentary on the malleability of the novel in terms of generic markers and its commitment to defying the classifications discussed at the beginning of this chapter. As such, *Indian Killer* is not just an example of any literary form, much like the "creatures of native literature are seldom mere representations of animals in nature or culture, wild, domestic, generic, or otherwise".[36] Rather, the text is de- and reconstructed in a process of infinite literary formation; it is detective in the sense that it uncovers colonial crimes and gives readers the chance to reconsider generic and ethnic stereotypes as well as learned beliefs about linear timelines, spatial singularity, and human–animal binaries. Vizenor observes that authored animals in many Native stories "are *real* as tropes of imagination";[37] they speak to a unification of all living organisms with similar rights to freedom and (political) agency. In *Indian Killer*, the owl likewise becomes a symbol of endless cultural creation and of a need for equal recognition of humans and other animals, allowing scenarios in which all colonial creatures join in a sacred dance of liberation.

"Remember, Human Beings Don't Control Anything. Spirits Do": Trickster Discourse in Adrian C. Louis's *Skins*

In contrast to Sherman Alexie's *Indian Killer*, Adrian C. Louis's novel *Skins* has not received much scholarly attention. Readings of the text's movie adaptation, directed by Cheyenne filmmaker Chris Eyre, have focused on the notion of the "Vanishing Indian",[38] Native spectatorship,[39] and Lakota masculinity,[40] discussing the story's narrative structure as well as its Lakota trickster figure, Iktomi, who shows up in the form of a spider. Peter Bayers, for example, argues that *Skins* is concerned with an "investigation into 'real' world questions about what Lakota masculinity can and should be in the early twenty-first century", relating this idea to the fact that Iktomi "generally represents everything a Lakota should not embody and, in this case, what a Lakota man should not embody".[41] My subsequent reading of *Skins* as a detective text that engages specifically in what

Gerald Vizenor calls "trickster discourse" follows these ideas by contextualizing the roles of the tribal investigator as protagonist and the spider as trickster as well as the historical and contemporary circumstances of colonial repercussions for the Lakota within the earlier outline of Indigenous detective writing.

While *Skins* confronts the reader with a main character whose job description reads criminal investigator for the Pine Ridge Reservation Public Safety Department, Lieutenant Rudy Yellow Shirt's investigative efforts are not the central concern of the novel. Like Alexie, Louis eschews straightforward cause-and-effect structures and rather invests in storytelling techniques that cut across any linear conventions, chronological patterns, or one-dimensional personae to emphasize the cyclic dimension of "the sacred hoop"[42] of Indigenous existences. Rudy's story is told in thirty-six parts, several of which jump back and forth between past and present, and between the physical and spiritual worlds, mid-chapter, without any textual indication, not allowing the reader to consume the narrative without piecing the temporal and spatial puzzle together themselves. While readers of *Indian Killer* are expected to serve as surrogate detectives in search of the eponymous murderer,[43] *Skins* presents an entirely different central premise: "Rudy sometimes believed that the only solution to violent problems was in more violent actions. [...] His guess was that maybe most men all over the world were this way. Mankind had always been a most violent animal" (35). Like John Smith and the Indian Killer, Louis's main character is seemingly forced into inevitably violent acts; in this case, however, Rudy's need for justice stems not so much from an enraged desire for vengeance, but from his desperation as an Indigenous individual caught in a whirlwind of crime, violence, poverty, and alcoholism not even his status as a tribal cop can overcome. Rudy understands and laments that "[i]t was bad enough that Indians got dumped on the reservations, but what was worse was that they learned to oppress themselves" (35). He implies that domination and harassment from the outside, mainly the federal government, has led to an inescapable state of paralysis for many Indigenous peoples from which they cannot escape. The Lakota on the Pine Ridge Reservation are caught, as becomes apparent in the novel, in a downward spiral triggered by disease, unemployment, and financial ruin, which, without the help of that same government, seems unmendable. At the same time, however, many men on the reservation reject even the idea of considering federal aid, leaving them without a possibility to move forward.

At this point, *Skins* emerges as a narrative defined by hopelessness and calls for an agent of liberation, who arrives in the form of the trickster spider Iktomi from another realm. As Alexandra Ganser observes, the Native trickster serves "as a source of energy that constantly disrupts the dominant simulations of Native America", as the kind of discursive sign "set free by cultural difference and moments of transdifference" between humans and animals, Natives and non-Natives.[44] Because the trickster "is never tragic", as Gerald Vizenor has famously theorized, but rather undoes tragedy through impulses of self-reflection and self-liberation, it cuts across states of helplessness, hardship, and erasure and thereby ensures the presence of Native stories and people.[45] For Vizenor, the trickster is a "comic and communal sign" in tribal cultures: it is collective as well as comprehensive and represents "imagination (and) liberation." Rather than just a being in animal form, it also signifies "a 'doing' in narrative points of view, and outside the imposed structures" of political ideologies and literary form.[46] Trickster discourse does not allow absolutes or binaries of good or evil, black or white, physical or metaphysical; rather, it is a "*holotrope*", a "whole figuration", and thus promotes wholeness of existence and discourse within and beyond respective texts, realities, and realms.[47] The trickster, especially as an animal figure, also pushes for transgression on multiple levels: while in many of Vizenor's works, for example, characters transform from human into animal states and back at will, in *Skins*, the trickster arrives in animal form and uses its powers to bring the human protagonist to a point of self-reflection. This process merges the roles of detective and (vigilante) villain in the same character, whereby Rudy's opportunity for criminal investigation is, at first, undermined yet ultimately steered in a productive and resourceful direction.

The novel thus proposes a detective plot in which Iktomi, the Lakota trickster, who first appears to a 12-year-old Rudy in the form of a black widow spider and bites him, emerges as a mediator of spiritual transformations in the protagonist and ultimately enables his journey towards self-discovery and self-acceptance. Almost thirty years after Rudy first encounters the "shapeshifter associated with spiders and with language and innovation",[48] he finds himself feeling like a parasitic animal—"a tick trying to secure a foothold on the mangy and rabid cur called America" (37)—constantly on police patrol along the margins between his Native reservation, where alcohol is prohibited, and the neighboring US town of Whiteclay, his alcoholic brother's prime supplier of beer. When Rudy trips and hits his head on a rock, "his vigilante other self, the one he called the

'Avenging Warrior' [is] born" (39), causing him to act out in vicious ways against his own people. Rudy initially considers Iktomi an "evil spirit" (48), but, according to Vizenor, "the trickster bears no evil" for "evil would silence the comic *holotropes*" of the trickster's figuration[49] as well as the animal's good-natured power to assist Rudy in (re)creating himself. While Iktomi is responsible in the novel for the protagonist's alter ego and outbursts of violence, the trickster's plans show signs of educational purpose: Rudy must acknowledge that like in "the old days, it was the family, the band, the tribe that came first" (21); that to "forget the animals, the sun, the waters, the rocks, and all the other living spirits was especially bad for an Indian" (45), and that he must reconnect with his Native environment, the spirit world, and the sacred hoop to become whole. Through Iktomi, Rudy, who "had felt and lived extreme violence, but also alienation, and rootlessness most of his life" (37), must learn to look beyond the physical dimension—where his repetitive patrolling activities offer no creative release—to a plane where the (dis)order proposed by the trickster enables him to reenter into conversation with himself as both human and animal. While the spider first spurs on Rudy's vigilante fantasies, Iktomi's shifting form from an animal to a sacred spirit guide ultimately speaks to the kind of Native sovereignty that allows the protagonist to change shapes and find a new way of life among his tribe—beyond the sorrows of the reservation, the violent tendencies of the "Avenging Warrior", and his hopeless efforts as a policeman. When his brother Mogie is diagnosed with liver cirrhosis, Rudy realizes "that it was getting a little lame to blame all his crazy thoughts on *Iktomi*" (178). After all, "in trickster discourse, the trickster unties the *hypotragedies* imposed on tribal narratives" from the imperial outside and uses its all-encompassing mythical powers to inspire spiritual progress and growth, not adversity or regress.[50]

After engaging in a sacred sweat lodge ceremony, Rudy "felt he had discarded *Iktomi*'s hold on him. And, finally, he hoped he had thrown away his alter ego, the 'Avenging Warrior.' Rudy Yellow Shirt felt a huge flood of spiritual relief. For once, he felt almost complete. Almost, but not quite. Something was still missing" (305). In order to complete the sacred hoop of his human–animal (trans)figuration and to free himself from the oppressive chains of colonial violence, Rudy must dip into the trickster's healing powers at will and in a creative way. In a comic "statement to white America" (189) following his brother's death, Rudy fulfils his promise to Mogie to vandalize Mount Rushmore by "giving George [Washington] a bloody nose. Pour five gallons of red paint down his face"

(188). As he ascends the hill, the bucket of paint the color of blood in hand, he looks to the sky and sees "a bright star twinkling. Maybe it was Mogie" (306), as Louis suggests a few chapters earlier in an exploration of Mogie's move from the physical to the spiritual plane: "Mogie Yellow Shirt was walking among the stars [and] in the distance he could see a beautiful green valley filled with people. Even from far away, he could tell they were smiling and beckoning him forward" (282).

Like Rudy after the healing ceremony, Mogie is able to escape the shackles of his cultural prison and enter the spirit world in full health to create a new life story beyond the earthly world. Like John Smith and the Indian Killer, Mogie is also imagined after death, in an ongoing process of creation that transcends all humanity, mirrored in the "wide path of red paint stretched down Washington's face, from his forehead to his chin" (306), the novel's final act of resistance to Native erasure. Reading *Skins* through trickster discourse as a contemporary iteration of a tribal creation story of the Lakota, on the one hand, and the Yellow Shirt brothers, on the other hand, suggests that Mogie and the spider Iktomi equally emerge as boundary crossers—between the physical and metaphysical dimensions, between human and animal form—and are thus able to break through the invisible seam between the planes one last time to complete the circle of life "as creators and markers of the cosmological realms".[51] Their ultimate act of vandalism against the discovery doctrine and centuries of hegemonic dominance, represented by the stone presidents, points towards a necessary re-narrating of Indigenous history from the point of view of the animal-trickster as creator. Iktomi inspires Rudy's acceptance of his own spiritual strengths and human shortcomings and thus preempts his role as a detective in search of earthly answers: the case's resolution lies neither in the violent pursuits of the Avenging Warrior nor in the professional success of Rudy as criminal investigator. Rather, the powers of the trickster liberate Rudy from his self-ascribed responsibilities as the sole hero of the Lakota and allow the text to break free from generic constraints of detective fiction and from expectations of human superiority over (other) animals. Like *Indian Killer*, *Skins* hence rejects any absolute figurations of its genre or protagonist and favours the hybridity at the core of all texts, cultures, and species. Rudy's transformation comes full circle when he realizes that "human beings don't control anything. Spirits do" (159). His role as a legal agent is thus put "on hold while the concept of creation—creation of the world, creation of the story, creation of any and everything—is emphasised",[52] putting Louis's novel into fruitful conversation with Alexie's in their emphasis of the critical synergies between (authored) humans, animals, Natives, non-Natives, spirits, and tricksters. Yet again, detection signifies a (re)discovery of tradition and identity through sacred creation in *Skins*: By infusing the investigator, the alleged figure of justice and reason, with both the chaos and the healing powers of the trickster, Louis emphasizes the fluidity of Indigenous existence and the importance of considering all realms and beings as parts of a comprehensive whole that is subject to continuous change. The novel emerges as detective through its uncovering of the inextricable link between the injustices resulting from colonial epistemologies and the imperative for Native self-help in this context: where law enforcement has its limits, the animal-trickster shows endless possibilities.

Conclusion: We Will Always Be Animals

Hudson claims that many "traditional Native American animal stories reinforce cultural values while also portraying how we should behave toward other species".[53] Reading Sherman Alexie's *Indian Killer* and Adrian C. Louis's *Skins* in dialogue with Native understandings of animal politics, ethics, and representations has shown that, in both novels, binaries and boundaries of existential and textual form no longer hold and that authored animals liberate tribal peoples and their narratives from the Western philosophies of representation. Vizenor describes these Western constraints as "tragic modes in translations and imposed histories", which only imagination and ceremonial healing in storytelling can undo.[54] In this sense, the creative discourses in *Indian Killer* and *Skins* ensure that "the tragic mode is in ruin" and that the allegedly bleak, vanishing realities of Indigenous peoples become enlivened with the spiritual significations of tricksterism and animal mythology.[55]

While Alexie's text provides a generally pessimistic outlook on the United States without any chance for resolution, the very absence of closure, a central investigative figure and plot mark the novel's potential for creation. Lydia R. Cooper suggests that "the small gestures [in the novel], like a kind hand and a shared sandwich, stand out in bold relief, the possible sacraments and rites of a new symbolism that affirms life".[56] What becomes even more prominent at the end of the story, however, is that "Indians are dancing now, and I don't think they're going to stop" (Alexie 418). This implies that potential Indian Killers can be re-birthed in endless cycles, but also that Native peoples are finding more ways for anti-colonial rebellion and productive re-creation of their histories through sacred ceremonies and storytelling. The revolutionary owl dance, a metaphor of the Ghost Dance, thus serves as *Indian Killer*'s ultimate tribal creation story: it enables John Smith to move on to the (next) metaphysical dimension, where humans and/as animals coexist in fruitful symbiosis and in constant spiritual relation with the earthly world. The novel's final words, "The tree grows heavy with owls" (420), additionally offer an alternative view of the killer as creator of a critical ecological awareness in the reader, beyond the textual boundaries of the narrative.

Louis's novel provides a similar focus on creation and the re-creation of history through the liberating effects of the tribal trickster. Iktomi, the Lakota shapeshifter and animal spirit guide, infiltrates Rudy's consciousness and teaches him that a deeper engagement with the beliefs and practices of his people promotes sustenance for creation and that "empathies fostered by such affinities inform social action".[57] Rudy's role as a tribal cop is de-emphasized—first in favour of his vigilante self, then in support of the importance of his human–animal hybridity—highlighting the nature of identities constantly in flux between varying poles. "*Iktomi* was an Indian giver" (Louis 301), the novel emphasizes, and, in this sense, offers both Rudy and readers a view of (tribal) worlds in which the alleged contradictions between Native and non-Native peoples, humans and animals as well as earthly and spiritual beings merge to create a comprehensive whole of existence that can be reinvented at any given point in time and space. While on the surface Rudy emerges as the novel's central criminal and thus erases his significance as an embodiment of the law, a closer look at his character in dialogue with trickster discourse shows that, like Iktomi, Rudy's shapeshifting qualities make it possible to exist as more than one entity and on several levels of consciousness.

In *Indian Killer* and *Skins*, animals infuse the texts and their protagonists with energy drawn from the sacred hoop of Indigenous existences, which are re-created in continuous motion and allow no rigid lines between species, races, and literary forms. Both texts eschew expected notions of detective fiction as imagined by Western epistemologies; they follow, instead, what I read as a Native American tradition of detective narratives wherein Columbian discovery and hegemonic misrepresentation are overhauled by tribal self-discovery and self-representation.[58] The fact that the novels *are* detective in an active sense of motion allows them to represent a continuance not only of Indigenous existences and agents but also, and especially, of traditions of storytelling that enter new literary discourses to map out, on the one hand, the intersections between the physical and the metaphysical as well as between the human and/as the animal, and allow, on the other hand, processes of (self)detection to cross between these different realms and bodies.

Alexie's and Louis's authored animals counter the dominant tragic representations of Native peoples as colonial creatures. They (re)create visions of worlds in which we are *all* animals and will forever soar beyond the horizons of colonial imagination.

Notes

1. Hudson, "Political Representation", 230.
2. Hudson, "Political Representation", 230.
3. Woodward and McHugh, "Introduction", 5.
4. Aftandilian, "Native American Theology", 195.
5. Lynch and Roberts, *Native American Mythology,* 5.
6. Anderson, "Animality", 3.
7. See Cox, "Native American Detective Fiction and Settler Colonialism" (2017) and "Crimes against Indigeneity: The Politics of Native American Detective Fiction" (2019); Stoecklein, *Native American Mystery Writing: Indigenous Investigations* (2019); and Hauke, "'Congress has never heard a voice like mine': Law, Legal Fictions and National Legal Culture in Native American Detective Fiction" (2020).
8. Cf. Berkhofer, *The White Man's Indian.*
9. Examples include *The Vanishing American* (1925), *A Man Called Horse* (1970), *Dances with Wolves* (1990), *Geronimo: An American Legend* (1993), or Disney's *Pocahontas* (1995).
10. Giles, *Spaces,* 143.
11. Berglund, *Cannibal Fictions,* 160.
12. The extensive corpus of this genre includes, for instance, Gerald Vizenor's *The Heirs of Columbus* (1991), Carole LaFavor's *Along the Journey River: A Mystery* (1996), Mardi Oakley Medawar's *Death at Rainy Mountain* (1996), and Stephen Graham Jones's *All the Beautiful Sinners* (2003), Tom Holm's *The Osage Rose* (2008), and Sara Sue Hoklotubbe's Sadie Walela series (2003–2018).
13. For a more in-depth discussion of my understanding of Native American detective fiction and the adjectival meaning of detect*ive*, cf. Hauke, "Law, Legal Fictions, and National Legal Culture".
14. Aftandilian, "Native American Theology", 195.
15. Aftandilian, "Native American Theology", 192.
16. Wolfe, *Before the Law,* 43.

17. Alexie, *Indian Killer*, 48.
18. Payne, "Border Crossings", 194.
19. Vizenor, "Authored Animals", 662.
20. Vizenor, "Authored Animals", 675.
21. Vizenor, "Authored Animals", 675.
22. James, "Captivity", 172.
23. Burnham, "Indigenous Gothic," 6.
24. Giles, *Spaces*, 128, 142.
25. Browne, *Murder*, 27.
26. Giles, *Spaces*, 135.
27. Giles, *Spaces*, 128.
28. Daniel Grassian cites an interview with Sherman Alexie in which the author explains that *Indian Killer* "was a response to the literary movement where a lot of non-Indian writers [were] writing Indian books. Non-Indian authors enjoy a success that is not determined or critiqued by American Indians. So I want to make sure they're aware of an Indian critical response to their work" (*Understanding*, 104). Alexie's rage seeps out of the novel at every instance and thus defines the criminal atmosphere in his fictionalized version of Seattle through "a critique" of what Lydia R. Cooper calls "violent atonement", highlighting the characters' "belief systems" as paradigms "used to justify violence against certain groups of people" ("Critique", 35).
29. Hudson, "Political Representation", 233.
30. James, "Captivity," 171.
31. Fee and Webb, *American Myths*, 667.
32. Cooper, "Critique", 39–40.
33. Giles, *Spaces*, 143.
34. Aftandilian, "Native American Theology", 197.
35. Woodward and McHugh, "Introduction", 2.
36. Vizenor, "Authored Animals", 678.
37. Vizenor, "Authored Animals", 678.
38. Cf. Feier, *Native American Cinema*.
39. Cf. Hearne, *Native Spectatorship*.
40. Cf. Bayers, "Spirituality".
41. Bayers, "Spirituality," 205, 196.
42. Louis, *Skins*, 21.
43. Merivale and Sweeney recognise such instances as "the detective's role as surrogate reader" (*Detecting Texts*, 2). In *Indian Killer*, the reader's—rather than a detective's—agency is foregrounded; hence my reverse understanding of this technique.
44. Ganser, "Transnational Trickster", 20, 24.
45. Vizenor, "Trickster Discourse", 284.

46. Vizenor, "Trickster Discourse", 285.
47. Vizenor, "Trickster Discourse", 282.
48. Hearne, *Native Spectatorship*, 344.
49. Vizenor, "Trickster Discourse", 285.
50. Vizenor, "Trickster Discourse", 283.
51. Aftandilian, "Native American Theology", 199.
52. Payne, "Border Crossings", 193–194.
53. Hudson, "Political Representation", 229.
54. Vizenor, "Trickster Discourse", 283.
55. Vizenor, "Trickster Discourse", 283.
56. Cooper, "Critique", 55.
57. Payne, "Border Crossings", 199.
58. Cf. Hauke, "Law", 171.

Works Cited

Aftandilian, Dave. "Toward a Native American Theology of Animals: Creek and Cherokee Perspectives." *Cross Currents* 61, no 2 (2011): 191–207.

Alexie, Sherman. *Indian Killer*. New York: Grove Press, 1996.

Anderson, Kay. "'The Beast Within': Race, Humanity, and Animality." *Environment and Planning D: Society and Space* 18 (2000): 301–320.

Berglund, Jeff. *Cannibal Fictions: American Explorations of Colonialism, Race, Gender, and Sexuality*. Madison: The University of Wisconsin Press, 2006.

Berkhofer, Robert. *The White Man's Indian: Images of the American Indian from Columbus to the Present*. New York: Vintage Books, 1979.

Bayers, Peter L. "Spirituality and the Reclamation of Lakota Masculinity in Chris Eyre's Skins (2002)." *American Indian Quarterly* 40, no. 3 (Summer 2016): 191–215.

Breitbach, Julia. "Rewriting Genre Fiction: The DreadfulWater Mysteries." In *Thomas King: Works and Impact*, edited by Eva Gruber, 84-97. Rochester, NY: Camden House, 2012.

Browne, Ray B. *Murder on the Reservation: American Indian Crime Fiction*. Madison, WI: The University of Wisconsin Press, 2004.

Burnham, Michelle. "Sherman Alexie's Indian Killer as Indigenous Gothic." In *Phantom Past, Indigenous Presence: Native Ghosts in North American Culture & History*, edited by Colleen E. Boyd and Coll Thrush, 3–25. Lincoln and London: University of Nebraska Press, 2011.

Cooper, Lydia R. "The Critique of Violent Atonement in Sherman Alexie's Indian Killer and David Treuer's The Hiawatha." *Studies in American Indian Literatures* 22, no 4 (2010): 32–57.

Cox, James H. "Crimes against Indigeneity: The Politics of Native American Detective Fiction." In *The Political Arrays of American Indian Literary History*, by James H. Cox, 177–206. Minneapolis: University of Minnesota Press, 2019.

Cox, James H. "Native American Detective Fiction and Settler Colonialism." In *A History of American Crime Fiction*, edited by Chris Razkowski, 250–262. Cambridge: Cambridge University Press, 2017.

Fee, Christopher R., and Jeffree B. Webb, eds. *American Myths, Legends and Tall Tales. An Encyclopedia of American Folklore*. Santa Barbara and Denver: ABC-Clio, 2016.

Feier, Johanna. *We Never Hunted Buffalo: The Emergence of Native American Cinema*. Münster: LIT, 2011.

Ganser, Alexandra. "Gerald Vizenor: Transnational Trickster of Theory." In *Native American Survivance, Memory, and Futurity: The Gerald Vizenor Continuum*, edited by Birgit Däwes and Alexandra Hauke, 19–33. New York: Routledge, 2017.

Giles, James. *The Spaces of Violence*. Tuscaloosa, AL: The University of Alabama Press, 2006.

Grassian, Daniel. *Understanding Sherman Alexie*. Columbia: University of South Carolina Press, 2005.

Hauke, Alexandra. "'Congress has never heard a voice like mine': Law, Legal Fictions, and National Legal Culture in Native American Detective Fiction." In *Crime Fiction and National Identities in the Global Age*, edited by Julie H. Kim, 161–187. Jefferson: McFarland & Company, 2020.

Hearne, Joanna. *Native Spectatorship and the Politics of Recognition in Skins and Smoke Signals*. New York: SUNY Press, 2012.

Hudson, Brian K. "A Seat at the Table: Political Representation for Animals." In *The Routledge Companion to Native American Literature*, edited by Deborah L. Madsen, 229–237. Oxon and New York: Routledge, 2016.

James, Meredith. "'Indians Do Not Live in Cities, They Only Reside There': Captivity and the Urban Wilderness in Indian Killer." In *Sherman Alexie: A Collection of Critical Essays*, edited by Jan Roush and Jeff Berglund, 171–185. Salt Lake City: The University of Utah Press, 2010.

Louis, Adrian C. *Skins*. Granite Falls, MN: Ellis Press, 2002.

Lynch, Patricia Ann, and Jeremy Roberts, eds. *Native American Mythology A–Z*. New York: Chelsea House, 2004.

Merivale, Patricia, and Susan Elizabeth Sweeney. *Detecting Texts: The Metaphysical Detective Story from Poe to Postmodernism*. Philadelphia: University of Pennsylvania Press, 1999.

Payne, Daniel G. "Border Crossings: Animals, Tricksters and Shape-Shifters in Modern Native American Fiction." In *Indigenous Creatures, Native Knowledges, and the Arts: Animal Studies in Modern Worlds* (Palgrave Studies in Animals and Literature), edited by Wendy Woodward and Susan McHugh, 185–204. Cham, Switzerland: Palgrave Macmillan, 2017.

Stoecklein, Mary. *Native American Mystery Writing: Indigenous Investigations.* Lanham et al.: Lexington Books, 2019.

Vizenor, Gerald. "Trickster Discourse." *American Indian Quarterly* 14, no 3 (1990): 277–287.

Vizenor, Gerald. "Authored Animals: Creature Tropes in Native American Fiction." *Social Research* 62, no 3 (1995): 661–683.

Wolfe, Cary. *Before the Law: Humans and other Animals in a Biopolitical Frame.* Chicago and London: The University of Chicago Press, 2013.

Woodward, Wendy, and Susan McHugh. "Introduction." In *Indigenous Creatures, Native Knowledges, and the Arts: Animal Studies in Modern Worlds* (Palgrave Studies in Animals and Literature), edited by Wendy Woodward and Susan McHugh, 1–9. Cham, Switzerland: Palgrave Macmillan, 2017.

FORMS

Aping the Classics: Terry Pratchett's Satirical Animals and Detective Fiction

Briony Frost

Predominantly known for his Discworld series, and sci-fi collaborations with writers such as Neil Gaiman and Stephen Baxter, Terry Pratchett may seem a strange choice for investigating the role of animals in detective fiction. His City Watch octet, however, belongs to the "blended genre [of] Fantasy-Detection".[1] Full of "swords, talking animals, [and] vampires",[2] the series uses frameworks and tropes from "Detective and Noir…crime fiction",[3] as well as hard-boiled and police procedural, to interrogate Western culture's priorities through "the grammar of punishment … what offences it recognises and punish[es], how it punishes them, the ways it attempts to justify punishment, the legitimising philosophies … and the complex rules governing who is allowed to punish who".[4] Pratchett's protagonist, Captain (later Commander) Sam Vimes, is a heavy drinking, down-on-his-luck, hard-boiled denizen. Insubordinate to higher authorities and with a "jaundiced view of clues"[5] that marks him out as the antithesis of Sherlock Holmes,[6] he polices Ankh-Morpork, the "oldest and

B. Frost (✉)
University of Plymouth, Plymouth, UK
e-mail: b.s.frost@bath.ac.uk

R. Hawthorn, J. Miller (eds.), *Animals in Detective Fiction*, Palgrave Studies in Animals and Literature,
https://doi.org/10.1007/978-3-031-09241-1_12

greatest and grubbiest of cities",[7] in cardboard-soled boots and exposes its corruption in fantasy's urban answer to the Wild West gunslingers.[8] Yet far from being isolated like Sam Spade and Philip Marlowe, and despite his own prejudices, he comes to command a "pluralist" force where "feminist, gay, non-white" and even non-human "voices are now heard".[9] By including among these voices not only humanoid beings such as dwarves, trolls, and elves, but also those belonging to non-human animals, Pratchett questions and challenges the vexed human/animal binary wherein "human" is "contingent upon its relation to 'the animal'", which results, like most categories of otherness, not in opposites but "codependent subjectivities".[10]

Pratchett's Watch series is thick with intertextual allusions to other detective stories from all genres, but in his exploration of inter-reliance of human and non-human animals it is his debt to the classics that stands out. *Guards, Guards!* (1989), the first of the Watch books, makes numerous playful and multifaceted allusions to the works of Edgar Allen Poe, Wilkie Collins, Arthur Conan Doyle, and Anna Katherine Green as Pratchett interrogates the permeable borders between human and non-human animals through the crime-scene encounters between them. Using critical perspectives on how humans construct and are constructed by animals, this chapter explores his satirical commentary on the anthropocentric thinking that "define[s us] as human in the face of the animal".[11] Of particular interest are the human detective's non-human behaviours as the Watch books contest perceptions, influenced by nineteenth-century atavistic discourses, that humans who behave like animals are criminals "acting automatically or under an insane and irresistible impulse to evil".[12] Instead, Pratchett suggests, it is animality within and beyond the human law's agent that lets him sustain his accountability and pursue legal justice.

Set in a city that bears more than passing resemblance to London between the 1642–48 Civil Wars (*Night Watch,* 2002) and the advent of railway (*Raising Steam, 2015*), Pratchett's Watch series exists in a generically hybridised space imagined prior to crime fiction's Golden Age where premodern jurisprudence governs. Not only may humans behave like animals and vice versa, but both may be punished accordingly. Pratchett's principal debt is to Poe, in part because of the latter's fondness for combining "the odd and unusual", "esoteric learning", and the "sciences of his day" mark him as surely as a science fiction pioneer[13] as his Monsieur C. Auguste Dupin identifies him as a grandfather of detective fiction.[14] Crucial for this chapter is that, as in Poe's own works, Pratchett yokes together genres where the boundaries between the human and animal in society are routinely questioned.

Emerging during the mid-nineteenth century, detective fiction's human/non-human animal relations are often inflected with the pseudo-scientific beliefs of the time when theories of criminality frequently made biological inheritance a scapegoat for deviance. Misreadings of Darwin's theory of the continuity between humans and animals, such as Cesare Lombroso's, which also drew upon the practice of physiognomy and phrenology, "opened up the disquieting possibility of regression" from a human to a bestial state or, worse, of being born "animal", lacking "absolutely every trace of shame and pity, [which] may go back far beyond the savage, even to the beasts themselves ... prov[ing] that the most horrible crimes have their origin in those animal instincts".[15] Detective fiction, thus, has a strong tradition of treating animals as "a foil against which to define the distinct [civilised] features of the human",[16] with works such as Poe's *Murders in the Rue Morgue* and Doyle's *The Hound of the Baskervilles* often (I shall later suggest, problematically) cited as exemplars.

Detective fiction has equally offered more positive models of human/non-human animal relations—from intelligent canine side-kicks, such as Tommie in *My Lady's Money* (1877); to emotionally supportive pets that provide much needed sanity during the trauma of a case, as Peppy and Mitch do for Sara Paretsky's V.I. Warshawski; and a whole host of contemporary animals-as-detectives—though all these tend to be constrained by their positioning of animals as liminal to human culture. Science fiction and its close cousin fantasy, however,

> offer a wider scope than ... most literature for enabling animal agency to become part of the quotidian world, as well as a space to attempt to grasp animals as beings in their own right rather than as beings defined through their place in human cultural systems.[17]

As Pratchett demonstrates through his use of the non-human animal as one among a number of minority voices now becoming audible, sci-fi and fantasy further disturb those divisions by "link[ing] the mistreatment of women, non-whites, and the working classes to the mistreatment of animals" or by representing "alien characters ... in terms that we typically associate with animals, raising questions about how we interact with living animals ... human/animal symbiosis, and animals as companions or fellow sentient beings".[18]

Where Pratchett pays homage to his ancestors he is often inviting a rereading of their depictions of animal-criminals and their place in the human world. His hat-tipping is both an honouring and a resisting of the classic traditions, which is spelt out most self-consciously in relation to the role of animals in detective fiction in *Night Watch* (2014). During his sixth case, Vimes travels back in time to ally with his youthful self, Young Sam, and confront a Lombroso-esque figure, wielding anthropological implements designed to detect criminal minds through cranial measurements in a collision of "*the mythological ethology* of the wild animal" and "the *anthropological mythology* of the primitive man".[19] Together, both Vimes succeed in overthrowing him, rejecting in full the legacy of nineteenth-century discourses aligning the animal and the criminal. This chapter concentrates, however, on earlier works in the series that enable Vimes to reach this stage in his investigation of the human/non-human animal borderland and which engage with some of the best-known non-human criminals of classic detective fiction—an orangutan, rats, and a (were)wolf. Re-presenting a detective's encounters with these beasts, Pratchett unsettles interpretations of his predecessors' depictions of them by opening up the cracks inherent in the essentialist discourse of human vs animal. In doing so, he forces his readers to consider the potential for animal instincts to lie at the heart of a "man of honour" who pursues justice in the "mean streets".[20]

I: "The Curious Incident of the Orang-utan in the Night-time"

Poe's "The Murders in the Rue Morgue" is largely credited as detective fiction's inaugurating work. Its central culprit—who slaughters two reclusive women at home in the titular street—is an "Ourang-Outang of the East Indian Islands" possessing a "wild ferocity", "superhuman" strength, "praeternatural … ability", and a voice "devoid of all…intelligible syllabification".[21] The investigation narrative identifies the non-human animal as acting "without motive" (402, 394). But the crime narrative discredits this. The murders are catalysed when a sailor discovers his orang-utan in his bedroom "attempting the operation of shaving" (407). The sailor, frightened by the animal's possession of a weapon, retaliates by grabbing a whip. The ape flees through a window, is startled by the screams of the occupants, and kills them. Critics such as Ed White and Elise Lemire have read this as an "allegory for the doctrine of black animality" and white anxieties interest in human/non-human primate similitude[22] and "emergent scientific discourses that threatened to locate *all* humans squarely within the domain of the animal".[23]

This mistaking of animal for human anticipates without authorising Lombroso's theory of the atavistic criminal. Instead of enabling humanness[24] to be defined against animalness, Poe, like Darwin, indicates that "animals possess some power of reasoning", since they "may constantly be seen to pause, deliberate, and resolve", as well as communicate both with each other and across species boundaries through gesture.[25] Poe stops short of fully crediting the ape with a conscience, but his representation of the creature's "fury", "dread", and "fear", as well as its being "[c]onscious of having deserved punishment", suggest that emotional and rational boundaries erected between human and animal are porous; the sailor runs this same gamut of emotions and understanding when confronting his part in the murders (409). Biological separations are further destabilised by "the analyst throw[ing] himself into the spirit of his opponent"[26] in an attempt to cross species boundaries, to "be" the ape despite differences of embodiment, activity, and sensory apparatus to solve the crime.[27] While police and earwitnesses flounder, because their understanding relies on reading actions and language within a human frame, Dupin, who is as "wild"[28] and isolated as the ape, "identif[ies] with the Ourang-Outang"[29] animalising himself to reread the crime text through a cross-species framework and close the investigation.

The potential for ontological human and non-human categories to dissolve into one another is further perpetuated by the ape's performance. The creature kills by literally "aping" the ritual of barbering, but slips from one practice commonly performed by trained barbers (shaving) to another (blood-letting) whilst fleeing the master's whip.[30] Human actions cause the crime: imitating and intimidating. The creature is not mistaken for human because, as Lombroso argues, some humans are still animals and therefore not human at all; rather, the ape's replication of and response to human behaviour aligns Poe's crime narrative with Darwin's cross-species empathy and tales that humanise animals. Although Poe's ape is not sentimentally humanised, when combined with his animalisation of the detective the crime text exposes the brutal animalisation of humans as one half of a binary, where the other recognises shared qualities of humanness. Poe's representation of both ape and man thus approaches what Ralph Acampora has termed "symphysis": an attempt to reach a common understanding based on shared embodied experience through which it is possible to arrive at "some comprehension of what it means to *be-with* other individuals of different yet related species".[31] A symphytic world does not manifest here—there is no moment of shared understanding between ape and detective, and the reader is distanced from both of their perspectives by the narrator—but the shadow of it lingers in the crime narrative, where it is picked up and brought into being through Pratchett's reimagining of another crime scene where ape and detective confront one another.

Pratchett's representation of an orang-utan in *Guards, Guards!* acknowledges the long history of the non-human animal's alterity but, more extensively and self-consciously than Poe, he uses it to challenge the human/non-human hierarchy. Disrupting the language of scientific racism infusing both classic and hardboiled detective fiction, the word "ape" is not only a noun for a 300lb primate but the term by which his orang-utan—the Librarian of the magical Unseen University—demands to be known. While "anthropoids" is accepted and Lady Sybil's reference to the Librarian as "my man" "let … pass without comment", Sergeant Colon agonisingly discovers that "'Ape' is all right, sir, but not the 'M' word … Just don't say monkey. Ohshit".[32] The Librarian's hostility to "monkey", while "ape" is acceptable and "man" tolerated, defamiliarises "ape" as a pejorative term for persons oppressed by the dominant social power and suggests an alternate species hierarchy in which the ape is superior. There are corollaries here with postcolonialism: writers of colour have historically inverted strategies that marginalise non-dominant human groups to categorise and marginalise hegemonic figures instead.[33] But to see Pratchett's focus as exclusively postcolonial overlooks his re-presentation of the non-human animal's consciousness and agency. Like Poe's ape, the Librarian sometimes performs human behaviour but, where the former's consciousness is only suggested through the detective, Pratchett makes us privy to the ape's inner thoughts and individual gaze. Explaining the central crime of *Guards, Guards!*—the theft of a book[34]—to the Watch's new recruit, Corporal Carrot, Pratchett uses the Librarian's perspective. He makes the reader party to the ape's exasperation with the human as he is forced to resort to using Charades, a game reliant on "theatrical" gestures (132), not because of his own limited capabilities but because the human struggles to participate in a mutual exchange with another species. As Carrot "sweat[s]" and fails to interpret the most basic of clues, the ape's frustration is noted in his dismissive thought: "Homo sapiens? You could keep it" (132, 133). Later, the Librarian attempts to produce an Orangutan/Human dictionary and eventually trains the Watch to realise "Ook" translates into complex English sentences. The Librarian thus advances Poe's representation of the limitations of using a human-only framework to read the world and reiterates Darwin's disbelief at the human "presumption in imagining … we understand the many complex contingencies on which the existence of each species depends".[35] Pratchett takes this sense of human arrogance and ignorance further, however, by positioning the reader as sharing the ape's point of view on the human detective, inviting us to "be" the ape ourselves, looking critically at our own species as if it were alien.

Poe's crime narrative hints the detective must have animal in him, allowing for slippage between human and non-human animal subjectivities and suggesting that traditionally hierarchised relations should be configured not as

$$\frac{\text{Human Animal}}{\text{Non-human Animal}}$$

but as parallel and interdependent identities (Human Animal/Non-human Animal). Pratchett, however, suggests that a subverted hierarchy

$$\frac{\text{Non-human Animal}}{\text{Human Animal}}$$

is equally valid through the eyes of the Librarian. Sworn-in as "Special Constable" and thus becoming a detective-protagonist himself, the Librarian's beat is literally and metaphorically above the Watch's—the "rooftops of the city"—and, unlike most Ankh-Morpork citizens, he "regards it as [his] duty to assist" (108, 156). In another satirical nod to Poe's ape, who rips open a window before criminally slaughtering the inhabitants, Pratchett's leans down from a rooftop to destroy a barred dungeon window (where Vimes is briefly imprisoned) and "the longest arms of the Law grab the astonished [Captain]" and pull him free to continue his pursuit of justice (347). Borrowing too Poe's detective-becoming-animal, the ape's strength is matched by Vimes'. Having discovered that the theft of a book has resulted in the summoning of a dragon that will destroy anyone in the city that resists its autocratic rule, Vimes is "angry enough to lift [and shake] 300lbs of orang-utan" and, with a "mammalian animal…noise", sets out to arrest the culprit (348, 350). Pratchett not only demolishes Poe's "intellectual man" and "strong man" categories,[36] but subsequent readings of them as reflecting respectively the human and non-human animal. Vimes' "superhuman" display of strength and animalistic exposition signify his "loyalty to his profession"[37] and reiterate what is implicit in Poe: that "animals are never fully animals and humans are never fully human but amalgamations of both".[38] The Librarian chooses to ally with, and occasionally behaves like, the humans of the Watch; Vimes finds him a better citizen than his own species and embraces his own animality to protect the City. As the Librarian himself puts it, "there [a]re times when an ape ha[s] to do what a man ha[s] to do".[39] Having reversed the human/non-human hierarchy, Pratchett therefore propels us towards Donna Haraway's conclusion: "Earth's beings are prehensile, opportunistic, ready to yoke unlikely partners into something new, something symbiogenetic".[40] He builds on this premise as the Watch becomes increasingly pluralist, suggesting that "coconstitutive companion species and coevolution are the rule, not the exception" (32)—for those on the right side of the law.

II: The Rats' Chamber and Animals as Metaphor

Before exploring the amalgamate nature of human and non-human animals, coevolving as detectives, the relationship of animals to criminality in *Guards, Guards!* needs further consideration. Two non-human creatures are central here. The first cannot straightforwardly be classed as either a human or a non-animal, but occupies a space that simultaneously belongs to and is rejected by both. This is the dragon. The dragon is both "myth", a creature summoned from the culturally diffuse and ephemeral space of lore and legend via the stolen book, and "real" in the sense that the local dragon expert, Lady Sybil, identifies Vimes' plaster cast of a footprint taken from the scene of the first crime as belonging to a biological organism, "*draco nobilis*" (130), that is now consigned in fossilised form to museums. Pratchett draws together here two of the three key theories of draconian genesis during Vimes' investigation of its city-wide coup, before moving us towards the third rationale for the existence of dragons: that the dragon is "a physical phenomenon of a nonbiological nature, or "symbolically equivalent to some psychological force".[41] By positioning the dragon as both real and not real, Pratchett leads his readers to a place from which we can reflect upon the relationship between human and non-human animals by encountering it through a process of defamiliarisation: the dragon is both and neither the human and the non-human animal; united in their confrontation of it, we as readers are forced to consider the human and non-human animal as allies against another creature—that which represents elements of each and can be accepted by neither.

Draco Nobilis usurps the city's Patrician and topsyturvies the pseudo-democratic social order. The dragon is a tyrant king, seeking to strip people of free will to serve its capitalist interests. In a biting critique of despotism and democracy, it swiftly learns its obsolescence: indoctrinated by persuasive rhetoric, people will sacrifice goods, homes, and even families and "come to believe it was their own idea" (295). The dragon, "*supposed* to be cruel, cunning, heartless, and terrible", flinches from the realisation that humans can manipulate communication until they "burn and torture and rip one another apart and call it morality" (296). The dragon is, the Librarian reveals to Vimes in another example of his potential superiority, a metaphor in the way that the non-human animals of Pratchett's world aren't. Dragons *"dwellyth in some Realm defined bye thee Fancie of the Wille, and, thus it myte bee thate whomsover calleth upon them, and giveth them theyre patheway unto thys worlde, calleth theyre Owne dragon of the Mind"* (395; italics original). They are invented, much like the atavistic criminal, to provide a shape for humans' own darkness (395). Nonetheless, while Poe's "motiveless" ape cannot be arrested nor his human keeper, Vimes leads us beyond the "limits the law imposes on notions of subjectivity" from the nineteenth century forward.[42] Echoing practices in parts of medieval Europe from "1280 until as late as 1750",[43] as well as the theories of nineteenth-century criminal anthropologists such as Havelock Ellis, Arthur Cleveland Hall, and Lombroso that deemed animals capable of criminality,[44] Vimes arrests the dragon in a symbolic gesture punishing a symbol of human evil. However, he releases it back into the wild margins of his world in favour of updating Poe's narrative to punish its primary human handler. He also calls attention to the anthropomorphic recruitment of animals as mere symbols for humanity and questions the ethics of what Steve Baker has called the "politics of animal picturing":[45] the linguistic and cultural conundrums encountered in imagining animals in contemporary culture when trying to recognise the sentience and identity of the animal as more than a non-human "other" against which humanness is defined.

If the dragon is a self-consciously transparent use of the non-human as scapegoat, the rat leads us into more complex territory. Rats as political metaphor have a long history. Among the most famous in detective fiction is Raymond Chandler's in *The Big Sleep*: Detective Philip Marlowe returns home to find his employer's daughter naked in his bed and, disgusted, recoils to his chessboard; he plays a knight, gloomily reflecting that "this wasn't a game for knights", while Carmen giggles "like rats behind the wainscoting" (72). The phrase encapsulates Chandler's world where "apparent respectability … masks a fundamental core of horror: corruption, perversity, death".[46] Like Chandler, Pratchett violates many of the rules set down by detective fiction writers and theorists S. S. Van Dine and Tzvetan Todorov; his world is inhabited by "many criminals", who "despite both logical explanation and physical carnage" cannot be "eliminated" and whose crimes "cannot be explained away as individual, non-social quirks or abnormalities".[47] Ankh-Morpork is akin to Michael Dibdin's Venice; "dwelling in the wainscoting of the country" is a ratking. A ratking is produced

> when too many rats live in too small a space under too much pressure. Their tails become entwined and the more they strain and stretch to free themselves the tighter grows the knot binding them until at last it becomes a solid mass of embedded tissue. … You wouldn't expect such a living contradiction to survive, would you? …[But] most … ratkings … are healthy and flourishing. … Each rat defends the interest of the others. The strength of one is the strength of all.[48]

The conference room at the Patrician's palace uses this cryptozoological phenomenon as metaphor for the city's politicians. The room is known as the Rats' Chamber, a multidirectional pun: on the German words for council chamber, "ratskammer", and town hall, "rathaus"; and an anagram of Star Chamber, the corrupt early modern English law court.[49] The Rats' Chamber features "a pattern of rats woven into the carpet" and a fresco of dancing rats, "their tails intertwining at the centre".[50] It suggests dirty political dealings: "after half an hour in that room, most people wanted a wash" (219).

Pratchett's rats-as-political-metaphor is most apparent via Havelock Vetinari. In *Guards, Guards!*, Vimes shares a cell with the Patrician who uses rats to spy on the city. Affected by "thaumic radiations" emitted from the Unseen University, they are "prod[ded] … towards minute analogues of human civilisation" (332). Robbed of his rule, Vetinari recreates "in miniature all the little rivalries, power struggles, and factions" (333). The rats are victims in the encounter, exposing the coercion of non-human animals into "fulfil[ling] human ideals … in anthropomorphic ways".[51] The rats, like their scorpion and snake cellmates, were removed from their natural habitats to serve the human animal's desire to torment other human animals. Like Poe's ape, they become hostile and violent towards one another: "it was sheer bedlam. The rats were getting the worst of it too" (332). In keeping with postmodern understandings of the non-human animal kingdom, which recognise that "nonhuman animals transmit behavioural information…[through] complex combination[s] of vocal, visual, tactile, and olfactory signs" that "when communicated across generations … may form a culture",[52] Pratchett's rats implicitly have a pre-existing culture and communication structure—which they later adapt to engage with Vetinari through gestures, such as nose twitching and "rathandling" and "listening" (334). But, in a continued analogue with other forms of alterity, what the rat is as a species and what individual rats are as "being[s] of 'unsubstitutable singularity'" is emptied out and replaced with a human idea of what rats signify; as humans historically have with all those deemed "other", the rats are thus constructed initially as "radically unlike ourselves in order to justify [human] behaviour towards them".[53]

Vetinari, seemingly benevolent compared to his predecessors, trains the rats to survive by behaving more like humans. Everyone in the city, from aristocratic assassins to common thieves, belongs to self-regulating guilds as, "if you must have crime, it might as well be organised".[54] The biggest criminals are the guild leaders, answerable only to Vetinari, whose Italian name alludes to his Mafia-boss role. Ratkings, in Ankh-Morpork, are not natural phenomena, as Pratchett confirms in young-adult work, *The Amazing Maurice and his Educated Rodents (2001)*. The villain, Spider, is a ratking, built by "cruel and inventive people…[with] too much time on their hands" (279). Here, anthropocentric thinking—"placing the human vision at the centre"—leads to active anthropomorphism—not just "seeing" but forcibly shaping "the world in our own image",[55] which Pratchett positions as, if not illegal, highly unethical.

Pratchett is not puritanical in his representation of animal ethics. Rats, for instance, are the staple diet of Ankh-Morpork dwarves, with both the Cut-Me-Own-Throat Dibbler, a human street vendor, and the dwarf Gimlet's Hole Food Delicatessen serving varieties of "rat-onna-stick" to the species. Echoes of Thai, Vietnamese, and Cantonese culture reverberate here which, alongside a complex and heterogeneous representation of the Dwarven faith, respectfully positions the species as culturally distinct from his humans (predominantly white and Western) who are the subject of his satire. Coexisting both easily and uneasily with the equally culturally and regionally diverse human world, the humans of Ankh-Morpork are lampooned by Pratchett for their fallacious anthropocentrism; as Fudge points out, all "human power over animals…undercuts humanity as a separate category".[56] Vimes notes Vetinari finds managing rats "easier than ruling Ankh, which had larger vermin who didn't have to use both hands to carry a knife" (333). Rather than subverting the human/non-human animal binary like the Librarian's presence, or serving being constructed as a metaphor for the darker aspects of humanness like the dragon, the rats demonstrate how anthropocentric world views "destroy *anthropos* as a category" and we "reduce ourselves to the thing we desire not to be".[57] Coevolution, then, is only positive where agency exists on both sides.

III: The Werewolf and the Monkey-Brain: Beyond the Binary

This chapter has considered Pratchett's reversal of the human/non-human animal binary and questioning of the ethics of using animals as metaphors. This final section returns to the Librarian and introduces the werewolf to move beyond the binary and into a world view that decentres the human. The orangutan's favourite form of identification is not "ape" but "Librarian". Changed by "magical accident" (12), he began his career as a human wizard and his transformation does not affect his identity within the University: "basically the same shape", he was "allowed to keep his job, which he was rather good at" (12) and "if someone ever reported that there was an orang-utan in the Library, the wizards would probably go and ask the Librarian if he'd seen it".[58] Historically, literary namelessness in fiction was a device reflecting the unimportance of traditionally marginalised human animal groups and non-human animals. But Claire Culleton, writing on James Joyce, notes that from the twentieth century onward nameless characters have become a "riposte…to the whole question of identity…no longer mute, no longer entirely powerless and certainly no longer marginal".[59]

A Hegelian influence is at work here. *The Phenomenology* observes that Adam's first act of naming animals meant animals ceased to exist as real creatures and became ideas: "naming immediately overturns what it names to turn it into something else".[60] Leslie Hill, quoted by Marcin Tereszewski, notes that "to supply any name is always tantamount to erasing a name" and "results in any name being a replacement, or a provisional name, of this original state of namelessness".[61] Ursula Le Guin adds that patriarchal power, language, and power over animals are linked: "in literature as in 'real life', women, children, and animals are the obscure matter upon which Civilisation erects itself, phallologically. That they are Other is (*vide* Lacan *et al.*) the foundation of language".[62] In "She Unnames Them" (1985), Le Guin's Adam's naming of the animals is reversed, rejecting the "conflati[on of] all members of a species to the generic species name, or ... into a singular non-human category".[63] As Derrida notes, this subverts humans' self-authorised "right and authority to [name] the living other" and so reclaims non-humans' "subjectivity and agency".[64]

In *Men at Arms* and *Fifth Elephant*, the werewolf-detective Angua reiterates this to distinguish between the wild wolves and domesticated dogs. "Wolves ... don't ... have names", she tells mutt-detective Gaspode, "Why should they? They know who they are, and they know who the rest of the pack are. It's all ... [s]mell and feel and shape. Wolves don't even have a word for wolves! ... Names are human things".[65] A therianthrope—the fullest manifestation of the human and non-human animal amalgamate—Angua struggles with her own identity throughout the Watch series, a battle encapsulated in the repetition of "w—" to refer to her in first appearances in *Men at Arms.* The curtailed naming of her marks her liminality within even Ankh-Morpork. Her character embodies "multiple anxieties of difference".[66] A "vegetarian...by day" and a "humanitarian by night",[67] she is rejected for her humanness by werewolves (especially her own aristocratic pack for whom serving in a human public police force is beneath them), for her animalistic hunger by the humans, and for "smell[ing] wrong" to the wolves.[68] Neither wholly woman, nor wolf, nor werewolf, and marginalised by all the groups to whom she belongs she is brought into the Watch as part of an initiative to "have a bit of representation from the minorit[ies]".[69] Her "w—"ness exemplifies the isolation experienced by those who occupy the intersectional spaces, possessing more than one minority characteristic.

Pratchett does not quite fully realise the problem of intersectionality—Angua is among the most recognisable and desirable of her kind, a "fair-haired", "light-eyed", and highly bred "lady lycanthrope" of the Victorian and Edwardian eras[70]—but through her alienation even in a unit that can embrace non-human animals, he draws attention to our tendency to deny commonality where a discourse of difference serves an agenda of privilege. Angua's viewpoint and the often uncomfortable space she occupies in the Watch alerts us to our own anthropodenial, our refusal to see similarities between fellow creatures and ourselves, which underpins the animalisation of other humans, such as the patriarchal proposition that women are closer to the animal kingdom and therefore "less evolved" than men. As with the Librarian, the reader is made aware of the flaws in essentialist discourses of animality by inviting us to share Angua's unique subjectivity. We see through her eyes "smells seared like brilliant lines of … coloured fire and … smoke", where "several hours" of the city's activity, individual identities, and emotions are visible "all in one go".[71] We share in her frustration at the lack of "proper words" to describe her wolf-self's multisensory experience and thus to convey it to her human companions, especially when the stakes in the case are high. We likewise cringe with her when her sex is subjected to the male gaze and the "w—" that is woman in her wolf is undermined simply for being not human and not male (42).

Although Angua does come to secure a place for herself in the pluralist Watch—one which encourages her to draw on her "abhumanity", making the most of her "morphic variability"[72] as both sniffer-dog and "her own handler"[73]—the process is gradual and erratic, haunted by twin discourses of anthropocentrism and sexism expressed by some of her senior colleagues, who read her as a "shaggy suffragette": "a cold-blooded threat to manhood and the status quo".[74] Contrastingly, the Librarian's place in society is much more consistent: his lack of name is not a marginalisation but an act of agency, a rejection of human identity; he excised his human name from university records to prevent his transformation being reversed. Embracing instead his privileged guardianship over books (repositories of language), the Librarian distances himself from the category of human, "othering" instead the traditionally dominant species and denying it authority over discourse. His recognition, and control, of the lethal potential of language to affect all strata of society, human and non-human, is encapsulated in his reaction to the theft of the book, which he equates to "genocide".[75] Pratchett's ape needs no blade to be dangerous; he controls access to the mightier outpourings of the pen—doing so as an indispensible associate of the law rather than an archive for all that humanity rejects about itself.

Furthermore, the Librarian as guardian of the book-as-weapon refocuses attention on the human animal's culpability in the crime. It is, he demonstrates, not only the dragon-summoner, at fault. It is the book's author, perpetuating the recruitment of animals as metaphors for human darkness. Neither "particularly holy" nor "noticeably evil", *Guards, Guards!* dragon-loreist's pathological normalcy is recognised as not only unethical but "bloody dangerous" (395). "Put that book somewhere safe", Vimes appeals to the Librarian, "and with it the book of law", which has been used to kill the dragon-summoner through a mistaken interpretation of another metaphor—"throw the book at him" (395, 388). The presence of both books throughout the narrative foregrounds the importance of "reading and writing" in the solving of crimes—a trait characteristic of most classical detective novels and canonised by critics such as Todorov, and Van Dine[76]—but in Pratchett how ordinary humans read the world and attempt to secure their place in it is where true crimes find their origin. Displacing animals as a biological origin of criminality, the exchange implies that it is humans' reading of the world that needs to be retheorised by criminologists and rehabilitated in dialogue with, not with its worst elements projected onto, the non-human animal to remain safe.

The most dangerous creatures in the Discworld are always human, with Pratchett taking this to its logical extreme in *Thud*, where the integrity of his detective is most starkly called into question. Possessed by the "Summoning Dark", an entity that brings out the worst in humans, Vimes comes close to surrendering to his own innate criminality. Confronting this text's nemeses, who are "cowering, already surrendering", Vimes "rais[es] the weapon above his head" (396). As he wrestles with his conscience,

> A werewolf landed on his back.
>
> Angua drooled. The hair along her spine stood out like a saw blade. Her lips curled back like a wave. Her growl was from the back of a haunted cave. All together these told the brain of anything monkey-shaped that movement meant death. And that stillness, while it also meant death, didn't mean immediate, *this-actual-second* death, and was therefore the smart monkey option (397).

Vimes' regression to a state of animality—"the smart monkey option"—is the moment his conscience triumphs over his impulse to evil. Angua's most dangerous shape, unlike Vimes' own, obeys her will. Her (controlled) bloodlust checks his (uncontrolled) lust for revenge by reminding him of his animal state. Once again, morality is positioned as belonging more to the animal than to the man as Vimes is startled into recovering his code of honour and remains loyal to his profession. The Summoning Dark is exorcised and a peaceful, legal, progressive solution found for the case. This episode brings together much of Pratchett's commentary on the co-dependency of human and non-human animal subjectivities. Rather than simply sustaining a subversion of the human/non-human categories, Pratchett's animals-as-detectives and detective-as-animal encourage us to dispense with the binary. Just as Angua learns to embrace her morphic mutability to become a respected asset to the Watch, Pratchett's detective-protagonist learns to embrace his "nonhuman" aspects in order to uphold the law when the human is too weak, weary, or wrathful to do so. Pratchett thus urges us as readers to recognise that coconstitutive relationships can be forged between human and non-human animals to moderate humanity's darkness.

IV: Conclusions

Pratchett's use of a human protagonist as principle narrator acknowledges the inevitable mediation through which non-human animals are understood. He does, however, often write from the point of view of his non-humans, emphasising the codependence of the species while acknowledging the limitations of human understanding of their non-human counterparts. Using the image of the ratking, Baker notes that "to try simply to tug free of the cultural constraints and stereotypes is to risk getting yet more firmly embedded in this living contradiction … to be free of the constraints and rules and influence of the ratking but powerless to change them … would be to have achieved nothing".[77] Pratchett does not escape the ratking, but nor does he accomplish nothing. As Simon Dannell has noted in his thesis on Pratchett's detective fiction "the established fantasy convention is that humans are the good characters, while the fantastic characters are the bad … Pratchett is subverting the traditional conceptions of fantasy literature whilst … retaining all the elements that make the story fantastic".[78]

Pratchett's generic hybrid suggests that, more than a human/non-human binary, "humans need fantasy to be human": "a place where the falling angel meets the rising ape".[79] The Watch series capitalises on the ways audiences are trained to read detective fiction to promote re-readings of the human relationship to the non-human world. Pratchett's Discworld does not place humans at its centre. Its humans place humans at its centre, but their encounters with the other species—demonstrated here through the ape, rats, and werewolf—make it clear that each of the other species also places itself at the centre of its world. The Watch books map a journey through and within a world similar to that which is posited in John Crowley's *Beasts* (1976), where "we are each of us living things, nothing but a consortium of other living things in a kind of continual parliamentary debate, dependent on each other, living on each other, no matter how ignorant we are of it; penetrating each other's lives" (9).

Within the Watch, and to a lesser extent within the Discworld more widely, human animals are part of a community with other beings that all possess individual subjectivity. There is still ignorance of this shared and codependent state, and Pratchett's world is not one that has freed itself from the discourses of alterity, especially speciesism. But it is a world in which his central human protagonist is conscious of his own ignorance, his prejudices and hostilities. As he confronts, revisits, revises, and rewrites them, he becomes more and more aware of his own flawed anthropocentricism and guides us as readers towards acceptance of the fellow subjectivity of the non-human animals.

Notes

1. Charles, et al., *The Readers' Advisory Guide,* 83. Pratchett now influences others, such as Jim Butcher's "cynical, wise-cracking" Harry Dresden, wizard-slash-detective protagonist of the Dresden Files, which are described as "a cross between Dashiell Hammett's Sam Spade and Continental Op mysteries and Terry Pratchett's wildly imaginative Discworld series" (83).
2. Pratchett, *A Slip of the Keyboard*, n.p.
3. Butler, *An Unofficial Companion,* n.p.
4. Bell, *Literature and Crime*, 53.
5. Pratchett, *Feet of Clay*, 205.
6. Pratchett's characterisation of Vimes borrows from Poe and Doyle the rationality of the detective, since early on in his first investigation Vimes observes of his unusual nemesis—a dragon—that solving crimes means accepting "once you've ruled out the impossible then whatever is left, however improbable, must be the truth" (154–55)—a near word-for-word echoing of Conan Doyle in *The Sign of the Four* (1980) and a recycling of Poe's ratiocination theory. However, in *Feet of Clay* (1996), Pratchett also

explicitly rejects the reclusive super-detectives of the classical period in a lengthy critique of

> the kind of person who'd take one look at another man and say in a lordly voice to his companion, "Ah, my dear sir, I can tell you nothing except that he is a left-handed stonemason who has spent some years in the merchant navy and has recently fallen on hard times," and then unroll a lot of supercilious commentary … when exactly the same comments could apply to a man who was wearing his old clothes because he'd been doing a spot of home bricklaying for a new barbecue pit, and had been tattooed once when he was drunk and seventeen and … got seasick on a wet pavement (205).

7. Pratchett, *Guards, Guards!*, 9.
8. Pratchett cites Clint Eastwood's Dirty Harry as a major influence on Vimes, echoing his "the law, you sons of bitches" quip and parodying the "are you feeling lucky" speech (*Men at Arms*, 354; *Guards, Guards!*,136).
9. McCracken, *Pulp*, 61.
10. Boggs, "Animals", 126.
11. Fudge, *Perceiving Animals*, 4.
12. Evans, *Capital Punishment*, 93.
13. Poe wrote seven proto-science fiction stories, including the thematically related tales "The Unparallelled Adventure of One Hans Pfaal" (1835), "The Balloon Hoax" (1844), and "Mellonta Tauta" (1840).
14. Broadly speaking, science and detective fictions are closely related, each relying on a system of rules including: "extrapolation … defined as the imaginative projecting of developments which might … be possible on the basis of present scientific knowledge", and being "able to account for happenings in [the] story by natural laws" (Olney, 416). While Pratchett does not "avoid the supernatural" (Olney, 416), his *Science of the Discworld* grounds his universe's laws in contemporary scientific understandings that restore plausibility to supernatural figures and concepts. For further evidence, see for example: C. Olney, "Edgar Allan Poe—Science Fiction Pioneer", *The Georgia Review* 12, no. 4, (Winter, 1958): 416–421; Harold Beaver, *The Science Fiction of Edgar Allan Poe*, Penguin, 1976; Burt Pollin, "Poe and Ray Bradbury, A Persistent Influence and Interest", *The Edgar Allan Poe Review* 6, no. 2 (Fall, 2005): 31–38; Rachel Haywood Ferreira, "The First Wave: Latin American Science Fiction Discovers its Roots", *Science Fiction Studies* 34 (2007): 432–462.
15. Lombroso, *Crime*.
16. Vint, *Animal Alterity*, 5.
17. Vint, *Animal Alterity*, 6.
18. Vint, "Animals in That Country", 177, 178.

19. Mazerallo, "Cesare Lombroso", 99; italics original.
20. Chandler, "Simple Art", 12.
21. Poe, "Murders", 403.
22. A number of popular primate displays were held in Philadelphia, where Poe was living when he wrote the text, such as "the huge, hairy, red ourang-outang that was exhibited before astonished crowds at the Masonic Hall in Philadelphia in July 1839" (Meyers, *Edgar Allan Poe,* 123). The creature had arrived from Africa aboard the ship Sabulda, an event reported in several local papers that all noted "its resemblance to the human form", which went beyond that of "any specimen that has been exhibited in this country" (*Pennsylvania Inquirer*, 1st July, 1839; *Public Ledger,* 1st July, 1839). See, for example: Haraway.
23. Peterson, "The Aping Apes," 157; italics original.
24. I borrow here Fudge's term, as it circumvents some of the difficulties encapsulated in the words "human" and "humanity" and "reveals the divisions which exist between being a human and possessing human qualities" (*Perceiving Animals*, 9–10).
25. Darwin, *Expression of the Emotions*, 77, 74.
26. Boggs, "Animals", 131.
27. Nagel, "What is it like", 438.
28. Poe, "Murders", 382.
29. Boggs, "Animals", 131.
30. Brunton, *Medicine Transformed*, 65.
31. Acampora, "Animal Philosophy", 12.
32. Pratchett, *Guards, Guards!*, 190.
33. Similarly, *in Guards, Guards!* the dragon makes an "almost human" use of "ape" as a derogatory term for all humans to assert its own superiority (295).
34. The theft, in a sly nod to Green's *The Leavenworth Case*, in which an aristocrat is killed in his own library, is declared, "worse than murder"!
35. Darwin, *Origin of the Species*, 202.
36. Poe, "Murders", 370.
37. Marling, *Los Angeles Detective Fiction*, n.p.
38. Mattfeld, *Becoming Centaur*, 8.
39. Pratchett, *Guards, Guards!*, 344.
40. Haraway, *Companion Species*, 32.
41. Blust, "Origin of Dragons," 522.
42. Boggs, "Animals", 126.
43. Beirne, "Use and Abuse", 6.
44. See: Piers Beirne, "The Use and Abuse of Animals in Criminology: A Brief History and Current Review", *Social Justice* 22, no. 9 (Spring, 1995): 5–31, esp. 6–9.

45. Baker, "Escaping the Ratking", 217.
46. Rabinowitz, "Rats", 231.
47. Rabinowitz, "Rats," 231. Unlike Chandler, however, Pratchett recalls the classic genre and disturbs the white/blue-collar divide by insisting that his *is* a world for knights, but ones that are created through merit rather than inheritance. Vimes' promotion to Commander of the Watch at the end of *Men at Arms* comes with a rise to knighthood that is coupled with a reopening of all the old Watch houses, affirmative action hiring, new departments of forensics, plainclothes, and surveillance, as well as pensions for the watchmen and their families, as a result of his remarkable and honourable endeavours to rid the city of crime (286–7). Knighthood is redefined in Pratchett's books as something modern and "open to all, regardless of species" (285) which, I demonstrate in the final section of this chapter, is achieved through the relationship between human and non-human animals that functions as an antidote to dark and dirty human authoritarianism.
48. Dibdin, *Ratking*, 76–7.
49. American poet, dramatist and biographer Edgar Lee Masters noted that, by the time of Charles I, its council "could inflict any punishment short of death", wielded its power arbitrarily, and "spread terrorism among those who were called to do constitutional acts" (*The New Star Chamber and Other Essays*, 12).
50. Pratchett, *Feet of Clay*, 219.
51. Fudge, *Perceiving Animals*, 4.
52. Jablonka and Lamb, *Evolution*, 205.
53. Vint, *Animal Alterity*, 13.
54. Pratchett, *Guards, Guards!*, 59.
55. Fudge, *Perceiving Animals*, 8.
56. Fudge, *Perceiving Animals*, 8.
57. Fudge, *Perceiving Animals*, 6; italics original.
58. Pratchett, *Night Watch*, 58.
59. Culleton, *Names and Naming*, 21.
60. Hegel, *The Phenomenology*, 332.
61. Terezewski, *Aesthetics of Failure*, 33.
62. Le Guin, "She Unnames Them", 9.
63. Vint, *Animal Alterity*, 23.
64. Jackson, 208; cited in Vint, *Animal Alterity*, 92.
65. Pratchett, *Men at Arms*, 228.
66. McKay and Miller, *Werewolves*, 1.
67. Pratchett, *Men at Arms*, 75.
68. Pratchett, *Fifth Elephant*, 152.
69. Pratchett, *Men at Arms*, 31.

70. Cinnias, "Wicked Women", 37.
71. Pratchett, *Men at Arms*, 122, 327.
72. Hurley, *The Gothic Body*, 3.
73. Pratchett, *Men at Arms*, 417.
74. Cinnias, "Wicked Women", 37.
75. Pratchett, *Guards, Guards!*, 131.
76. Huhn, "Detective as Reader", 451.
77. "Escaping the Ratking", 188.
78. "Blurring of Genres," n.p.
79. Pratchett, *Hogfather*, 74.

Works Cited

Acampora, Ralph R. "Animal Philosophy: Bioethics and Zoontology". In *A Cultural History of Animals in the Modern Age*, edited by R. Malamud, 139–161. Oxford: Berg, 2007.

Baker, Steve. "Escaping the Ratking: Strategic Images for Animal Rights". In *Picturing the Beast: Animals, Identity, and Representation*. 187–236. University of Illinois Press, 2001.

Beaver, Harold. *The Science Fiction of Edgar Allan Poe*. Penguin, 2006.

Beirne, Piers. "The Use and Abuse of Animals in Criminology: A Brief History and Current Review". *Social Justice* 22, no 9 (Spring, 1995): 5–31.

Bell, Ian A. *Literature and Crime in Augustan England*. London: Routledge, 1991.

Blust, Robert. 2000. "The Origin of Dragons". *Anthropos* 95, no 2. 519–536.

Boggs, Collen Glenney. "Animals and the Letter of the Law". In *Animalia Americana: Animal Representations and Biopolitical Subjectivity*. 109–132. Columbia University Press, 2013.

Brunton, D. *Medicine Transformed: Health, Disease, and Society in Europe, 1800–1930*. MUP: 2015.

Butler, Andrew M. *An Unofficial Companion to the Novels of Terry Pratchett*. Greenwood World, 2007.

Cininas, Jazmina. "Wicked Women and Shaggy Suffragettes: Lycanthropic Femmes Fatales in the Victorian and Edwardian Eras". In *Werewolves, Wolves, and the Gothic*, edited by R. McKay and J. Miller, 37–64. Cardiff: University of Wales Press, 2017.

Chandler, Raymond. 1939. *The Big Sleep*. Penguin, 2011a.

Chandler, Raymond. 1950. *The Simple Art of Murder*. Vintage Crime, 1988a.

Chandler, Raymond. 1939. *The Big Sleep*. Penguin, 2011b.

Chandler, Raymond. 1950. *The Simple Art of Murder*. Vintage Crime, 1988b.

Charles, J. C., C. Clark, and J. Hamilton-Selway. *The Readers' Advisory Guide to Mystery*. Chicago: American Library Association, 2012.

Coleridge, Samuel Taylor. *A Selection of Shakespearean Criticism*. Edited by R.A. Foakes. Bloomsbury, 1989.
Conan Doyle, Arthur. 1939. *The Adventures of Sherlock Holmes*. London: Penguin, 2007.
Crowley, John. *Beasts*. Garden City: Doubleday, 1976.
Dannell, Simon. "Terry Pratchett, the Watch, and the Blurring of Genre". (Unpublished diss., University of Lincoln, n.d.). https://www.lspace.org/books/analysis/simon-dannell.html. Accessed: 15/3/18.
Darwin, Charles. *On the Origins of the Species*, edited by D. Waummen. London: Sterling, 2008. 1872.
Darwin, Charles. *The Expression of Emotions in Man and Animals*. 1872. Stilwell: KS, 2005.
Dibdin, Michael. *Ratking*. Faber & Faber, 2011.
Evans, E.P. *The Criminal Prosecution and Capital Punishment of Animals*. Law Exchange, 1906.
Frank, Lawrence. "'The Murders in the Rue Morgue': Edgar Allan Poe's Evolutionary Reverie". *Nineteenth Century Literature* 50, no 2 (1995): 168–188.
Fudge, Erica. *Perceiving Animals: Humans and Beasts in Early Modern English Culture*. Illinois, 2002.
Green, Anna Katherine. 1878. *The Leavenworth Case*. Create Space, 2016.
Haraway, Donna. *The Companion Species Manifesto*. Paradigm, 2003.
Haywood Ferreira, R. "The First Wave: Latin American Science Fiction Discovers its Roots". *Science Fiction Studies*, 34 (2007): 432–462.
Huhn, Peter. "The Detective as Reader: Narrativity and Reading Concepts in Detective Fiction". *Swedish Modern Fiction Studies* 33, no. 3. (Fall, 1987): 451–466 https://muse.jhu.edu/article/244364 Accessed: 16/02/18.
Hurley, Kelly. *The Gothic Body: Sexuality, Materialism, and Degeneration at the Fin de Siècle*. Cambridge: Cambridge University Press, 2004.
Jablonka, Eva. and Marion J. Lamb. *Evolution in Four Dimensions: Genetic, Epigenetic, Behavioural, and Symbolic Variation in the History of Life*. Cambridge, MA.: 2006.
Lemire, Elise. "'The Murders in the Rue Morgue': Amalgamation Discourses and the Race Riots of 1838 in Poe's Philadephia". In *Romancing the Shadow: Poe and Race,* edited by J. G. Kennedy and L. Weissberg, 177–204. Oxford University Press, 2001.
Lombroso, Cesare. *Crime: Its Causes and Remedies*. 1911. Quoted in Lawrence Frank, *Victorian Detective Fiction and the Nature of Evidence*. Basingstoke: Palgrave, 2003.
McCracken, S. *Pulp: Reading Popular Fiction*. Manchester: Manchester University Press, 1998.
Masters, Edgar Lee. *The New Star Chamber and Other Essays*. Hammersmark Publishing, 1904.

Mattfeld, Monica. *Becoming Centaur: Eighteenth Century Masculinity and English Horsemanship*. Pennsylvania State University Press, 2017.

Mazarello, Paolo. "Cesare Lombroso: An Anthropologist Between Evolution and Degeneration". *Functional Neurology* 26, no 2 (2011): 97–101.

Nagel, Thomas. "What is it like to Be a Bat?" *Philosophical Review* 83, no 4 (1974): 435–450.

Olney, Clarke. "Edgar Allan Poe – Science Fiction Pioneer". *The Georgia Review* 12, no. 4 (Winter, 1958): 416–421. http://www.jstor.org/stable/41395580 12/1/18.

Peterson, Christopher. "The Aping Apes of Poe and Wright: Race, Animality, and Mimicry in 'The Murders in the Rue Morgue; and *Native Son*". *New Literary History* 41, no.1 (Winter, 2010): 151–171. http://www.jstor.org/stable/40666489 01/03/18.

Poe, Edgar Allan. "The Murders in the Rue Morgue". In *The Complete Works of Edgar Allan Poe,* edited by J. A. Harrison, 378–410. Vol 4. New York: Thomas Y. Crowell and Company, 1902.

Pollin, Burton. "Poe and Ray Bradbury: A Persistent Influence and Interest". *The Edgar Allan Poe Review* 6, no 2 (Fall, 2005): 31–38.

Pratchett, Terry. *Guards, Guards!* Victor Gollancz Ltd, 1989.

Pratchett, Terry. *Men at Arms.* Victor Gollancz Ltd, 1993.

Pratchett, Terry. *Feet of Clay.* Victor Gollancz Ltd, 1996.

Pratchett, Terry. *Hogfather.* Victor Gollancz Ltd, 1997.

Pratchett, Terry. *The Fifth Elephant*. Doubleday, 1999.

Pratchett, Terry. *The Amazing Maurice and his Educated Rodents.* Doubleday, 2001.

Pratchett, Terry. *Night Watch*. Corgi, 2014a.

Pratchett, Terry. *Thud*. Corgi, 2014b.

Pratchett, Terry. *A Slip of the Keyboard*. Anchor, 2015.

Rabinowitz, Peter J. "Rats Behind the Wainscoting: Politics, Convention, and Chandler's *The Big Sleep*". *Texas Studies in Literature and Language* 22, no. 2 (Summer, 1980): 224–245.

Todorov, Tzvetan. "The Typology of Detective Fiction". In *The Poetics of Prose*, edited and translated by Richard Howard, 42–52. Blackwell, 1977.

Van Dine, S.S. "Twenty Rules for Writing Detective Stories". *The American Magazine*. 1928.

Vint, Sheryll. "'The Animals in That Country': Science Fiction and Animal Studies". *Science Fiction Studies,* 35, no 2 (2008): 177–188.

Vint, Sheryll. *Animal Alterity: Science Fiction and the Question of the Animal.* Liverpool University Press, 2012.

Wilkie Collins. 1878. *My Lady's Money.* Createspace, 2017.

White, Ed. "The Ourang-Outang Situation". *College Literature* 30, no 3 (2003): 89–108.

Animal Image and Human Logos in Graphic Detective Fiction

Joseph Anderton

The combination of detective fiction, comic books and non-human animals is a perfect storm of lowbrow popularism. Edmund Wilson denounces detective novels as boring "rubbish"; comic books are commonly associated with "the madcap, the childish, the trivial" according to Thomas Doherty; and the animal, as Steve Baker notes, "is the sign of all that is taken not-very-seriously in contemporary culture".[1] Only relatively recently have detective fictions, comic books and non-human animals attracted sustained academic study, yet detective stories featuring animals have long been visually informed. Early examples include Daniel Vierge's and Aubrey Beardsley's drawings of a frenzied ape for Edgar Allan Poe's *The Murders in the Rue Morgue* (1884–94) and the bloodhound–mastiff cross with a luminous head in *The Hound of the Baskervilles* (1901–02). Detective fiction more broadly has a rich history of appearing in illustrated publications and featuring accompanying images, prompting Christophe

J. Anderton (✉)
Birmingham City University, Birmingham, UK
e-mail: Joseph.Anderton@bcu.ac.uk

R. Hawthorn, J. Miller (eds.), *Animals in Detective Fiction*, Palgrave Studies in Animals and Literature,
https://doi.org/10.1007/978-3-031-09241-1_13

Gelly to claim "detective fiction in general entertains a privileged relationship with the world of images and illustrations".[2]

Animal images have been used in graphic detective fiction at least since DC's Detective Comics series introduced Batman in 1939. The title of the first Batman story, "The Case of the Chemical Syndicate", frames the character as a detective in a crime tale, working both with and independent of Commissioner Gordon and the Gotham City Police Department. Batman is sometimes known as the "World's Greatest Detective", largely because the epithet was self-bestowed on the cover of *Superman's Girlfriend, Lois Lane* in the late 1960s. Batman's German Shepherd, Ace the Bat-Hound, aided the dark knight in his fight against crime in several issues from 1955 onwards, possibly owing to the popularity of the Lassie films in the 1940s and television series in the 1950s, which followed on from the success of early canine film stars such as Strongheart, Rin Tin Tin and Ace the Wonder Dog. The post–Second World War years also saw the arrival of animal as detective in DC's Detective Chimp (alias Bobo T. Chimpanzee) (1952–1959), a trained great ape turned anthropomorphic "amplified animal" who wears a deerstalker cap, receives the ability to talk to humans and later becomes an alcoholic. DC's Detective Chimp first appears in the comic *Rex the Wonder Dog* (1952–1959), featuring another highly intelligent detective animal that solves crimes. A drunken Detective Chimp later turns up alongside Batman in *Injustice: Gods Among Us* (2014). This pairing of the apparently bat-like man and human-like ape highlights two ways that images of animals cross with the human. Batman externalises his deep fear of bats by adopting visual elements and behaviours to represent a bat, including a mask with large protruding ears, a cape to act as gliding wings, hanging upside down and operating at night. Detective Chimp, on the other hand, is an example of an animal that has learned or acquired recognisably human, and some superhuman, characteristics. Batman is therefore wearing the animal image as part of his performative alter ego, in contrast to Detective Chimp, who experiences a transformed, anthropomorphised ontology. Both indicate how a human identity can underpin the appearances of non-human animals in comic books, either as an original or imposed self.[3]

This chapter will examine two contemporary comic book detective series featuring animal protagonists: the noir cat investigator in the *Blacksad* trilogy by Juan Diaz Canales and Juanjo Guarnido (2000), and the Victorian sleuth badger in *Grandville* by Bryan Talbot (2009). These two books continue a tradition of anthropomorphic animals in comics, harking back to the "funny animals" popularised by Disney Comics in the 1930s and Fawcett Comics in the 1940s, in which cute and lightly comedic characters such as Mickey Mouse and Hoppy the Marvel Bunny are heavily humanised. Leonard Rifas defines the funny animal genre as typically featuring "characters who combine animal faces with upright bodies that include hands, dressed (at least partially) in clothes, who converse with each other using language rather than animal sounds (for example, Donald Duck). These characters think and act more like people than like animals".[4] *Blacksad* and *Grandville* participate in this anthropomorphic gravitation to the human, but are also notable for evolving the use of animal images from wholesome, superficial funny animals towards a darker, edgier and more complex form, particularly in the way they address mature themes; apply noir and steam punk art styles; are satirically critical of human societies; and draw liberally from established detective fiction genres in a playful but deferential way.

If detective fiction is implicitly humanist owing to the restoration of order through the application of human mental acuity, the animal detectives I consider here might serve to parody the aggrandisement of the mind by having feline and musteline creatures perform these ingenious feats. Yet, as Jim Steranko observes in his introduction to *Blacksad*, "rather than animals who act like people, the creators' approach is predicated on people who resemble animals".[5] Hence, the image of the animal actually appears to obviate the non-human, with real animals receding twice over. The animal heads are caricatured, anthropomorphic versions of the non-human, easily understood as superficial disguises worn over the depth of human logic and reason, or *logos*. *Blacksad* and *Grandville* both evoke notions of animalistic surface appearances and human inner substance, which can map Cartesian duality onto these books in the form of animals as body/image and humans as mind/text. However, in conjunction with various disruptive references in these comic books, I argue that the comic book form itself unsettles the animal image and human logos hierarchy, found ostensibly in their anthropomorphic aesthetics and detective fiction genre. The complex interplay between image and text in graphic narratives includes "rupture, relationship and synthesis" according to W. J. T. Mitchell, and can therefore subtly inform a more refined understanding of the human–non-human animal nexus.[6] In this multifaceted model, the animal image, and indeed animality, is not a superficial reduction, but a complementary and necessary presence.

Theriocephalic Detectives: Animal Image in *Blacksad* and *Grandville*

Blacksad and *Grandville* follow in the tradition of anthropomorphism in visual and textual cultures, which Talbot describes as "ancient and universal", stretching back to the human–animal fusions in Palaeolithic cave paintings, Hindu and Egyptian gods, Aesop's fables and Cherokee Indian trickster tales.[7] Although anthropomorphism can take different forms and varying degrees, the characters in *Blacksad* and *Grandville* are physically hybrid, or bimorphic, combining bipedal human bodies with animal heads. Juan Diaz Canales and Juanjo Guarnido's John Blacksad is a private investigator in hard-boiled crime stories, such as the murder of an actress, a kidnapping by a racist political organisation, and an attempted assassination of a nuclear physicist, which take place in various American cities in the 1950s. He has a black and white cathead and hirsute human body, although at least one character in the series has a tail: the white fox in "Arctic Nation". Blacksad wears dark suits, dress shirts and trousers, a loose tie, formal shoes and a beige long coat. The creators cite 1930s noir as inspiration, such as Dashiel Hammett, James M. Cain and Raymond Chandler novels, as well as the Argentina-born duo Carlos Sampayo and Jose Munoz's *Alack Sinner* (1975), a classic graphic novel series rendered in a noir, monotone art style and featuring a hard-boiled American detective. The artist, Juanjo Guarnido, is particularly practised in rendering anthropomorphic characters; he worked at an animation studio on television shows such as *Tintin*, *Adventures of Sonic the Hedgehog* and *The Pink Panther*, as well as an animator with Disney, before joining up with Canales for *Blacksad*. Although Steranko suggests Guarnido's depictions of "emotional nuances and facial expressions" are "easily the equal of any Disney effort on record", it is patent that the artist is departing from the juvenile, innocent world of cartoon animation and bringing the anthropomorphic animal into the graphic narrative's potential for more complex social and historical themes, including the grittier, adult realm of crime, violence and sex.[8]

Bryan Talbot's Detective Inspector Archie LeBrock of Scotland Yard investigates unexplained murders in an alternate history of late-nineteenth-century Paris, or "Grandville" as it is called in the book. In this fantasy steampunk world, Britain lost the Napoleonic wars and is now named the Socialist Republic of Britain, having only recently achieved independence from the pervasive French Empire. LeBrock has a badger head and a formidable human body. His detective style is similarly hybrid as an observant, deductive Golden Age detective but with a brute twist. Talbot names Arthur Conan Doyle as an influence and it is first noticeable in LeBrock's clothing, mainly his greatcoat, bowler hat and umbrella, which are reminiscent of Holmes' frock coat, cloth cap and cane.[9] LeBrock's partially Holmesian detective style is introduced during the first investigation in Nutwood (an intertextual reference to Rupert the Bear's home village) when LeBrock exercises his impressive powers of observation and deduction in a classic "locked room" scenario that also alludes to Poe's Dupin (12–16). Since LeBrock works for the Metropolitan Police Service, he is not an "amateur" detective in the mould of Dupin or Holmes, but it is the extremity of the violence in *Grandville* that really sets him apart. Whereas Holmes carries a pistol in stories including *The Sign of Four* (1890) and *The Hound of the Baskervilles*, and is said to be a fine boxer in "The Adventure of the Yellow Face" (1893), LeBrock wields two pistols, is a highly capable fist fighter, often uses physical force in his clashes with criminal gangs and even appears demonic in a couple of panels.[10] He is therefore part Golden Age detective, part ultraviolent anti-hero.

While the French illustrator Albert Robida inspires Talbot's industrial technology and aesthetic designs, the series is named after one of Talbot's other key influences, the French caricaturist Jean Ignace Isidore Gérard, who worked under the pseudonym J. J. Grandville. He was known for "clothing his animals in the fashions of the day and giving them human airs, gestures, emotions, and thoughts" in his illustration series "Metamorphoses of the Day" (1829) and "Scenes from the Private and Public Lives of Animals" (1842).[11] In addition to precursors such as John Tenniel and Edward Lear, Melanie Hawthorne suggests that "Talbot pastiches the style of Beatrix Potter", which is true insofar as he also uses animal heads with humanoid bodies.[12] Yet Talbot's art style is not especially imitative and differs markedly from Potter, indeed from all of the above, with his sharp black ink, atmospheric tints, vivid blood sprays and kinetic blurred effects. His aesthetic owes more to Quentin Tarantino, who receives an acknowledgement from Talbot, than Potter (10).

Although Blacksad and LeBrock are ostensibly an American shorthaired cat and Eurasian badger, respectively, their relationships with the real animals are limited by the extent of their anthropomorphism. Steranko suggests that in their depictions of the animal characters in *Blacksad*, Canales and Guarnido are "generally unconcerned with their zoological differences; they are cast for their natures and personalities", which means utilising their "intrinsic qualities" and resulting in "overt symbolism. Or simple typecasting".[13] Cats have several actual and apocryphal attributes that lend themselves to representations as noir detectives. They are vigilant, stealthy, inquisitive hunters. They are considered nocturnal, being well adapted to hunt at night, although they are better described as crepuscular creatures, most active in the twilight hours. They are commonly recognised as solitary beings, although domestic and wild cats can and do live like prides. As strays, they are hardy, resourceful survivors, giving rise to the myth that cats have nine lives. These real, sometimes exaggerated, feline behaviours translate to an aloof, obscure, persistent detective character navigating the shadowy world of crime and corruption. In the human imagination, cats are also deemed particularly clever, guileful, deep or sophisticated animals, especially in comparison to dogs. These more dubious impositions on cats are congruous with the idea of a sharp, unorthodox, brooding and complex noir detective.

Talbot selects a badger for his robust detective because of two very different reputed characteristics: "ferocity" and "tenacity".[14] Unlike watchful cats, badgers have poor eyesight and can be caught completely unaware when occupied smelling the ground or digging for earthworms. They are viciously defensive and unrelenting combatants in these circumstances, which is perhaps why later in *Grandville* the news media refer to LeBrock as a "large animal, thought to be a bear", when badgers are actually from the weasel family (97). Despite their reputation for being pugnacious, badgers are largely elusive nocturnal animals; they are underground dwellers, avoid conflict and are seldom seen. Overall, these accurate descriptions of a badger's behaviour amount to a suitable emblem for a powerful, mysterious detective character. Other more subjective interpretations, representations and misconceptions also add to the badger–detective pairing. Talbot's annotations for *Grandville* reveal that he chooses a badger for its distinctive visual impact; the black and white facial markings resemble a mask and make them look "cooler" according to the author.[15] Given Talbot's interest in anthropomorphic precedents, it is likely that Mr Badger from Kenneth Grahame's *The Wind in the Willows* (1908), a character that Daniel Heath Justice describes as "a skilled fighter who challenges chaos with force when reason and more conventional mores are no longer effective", influences LeBrock.[16] Finally, badgers evoke the verb "badgering", meaning "to hound" or "persistently harass", which is also relevant to an inquisitive detective, although in truth the term derives from badgers being vilified and baited, so that historically badgers were the victims of badgering.[17]

The rest of the animal casting in *Blacksad* and *Grandville* shares the same essentialist, totemic identity logic, and notably the henchmen are powerful animals, like rhinos and bears, while the villains are cold-blooded reptiles, such as snakes and lizards. Suzanne Keen acknowledges that "the representation of particular animals can rarely be a neutral matter. [...] Thus any anthropomorphized representation of an animal either tacitly accepts or works against cultural pre-sets".[18] Canales and Talbot appropriate these cultural pre-sets as shorthand symbolism to establish recognisable character types for the narrative genre. They use the composite animal identities that have accreted from human knowledge, perceptions and figurations derived from real animals to communicate approximations of each anthropomorphic character.

However, if the process of anthropomorphism can imply that the non-human animal is an original or base identity that is then transformed by the attribution of human traits, this definition is misleading regarding Blacksad and LeBrock. It seems backwards to principally identify them as animals that have been humanised when so much of their animality has been effaced. They have cat and badger heads, which positions the animal as the front of their identity, but almost everything else about them is recognisably human. The characters move, behave and talk like people; they wear clothes; they live and work in familiar cities. It is therefore perhaps more appropriate to apply the term "theriocephalic", meaning "beast head", as it captures the animal to human ratio of their appearance.

Alternatively, Lisa Brown's use of the term "reverse anthropomorphism" is applicable to an extent, meaning a zoomorphic transformation from human to animal. She writes: "By using subtle animal imagery in drawings of humans, comic artists can infuse their characters with deeper meaning without ever having their characters say a word. This is a kind of reverse anthropomorphism, in which people take on the characteristics of animals".[19] The possibility that these protagonists are animalised humans, as opposed to humanised animals, is most evident visually in panels showing them in states of undress after sex to emphasise the contrast between the animal head and the naked human body concealed beneath their clothing.[20] Such examples indicate that the human underpins the characters' identities, despite the animal imagery. Indeed, having noted that in *Grandville* there is minimal attention to animal proportions when comparing species, Hawthorne asserts that "Talbot is less interested in depicting 'nature', despite the realism of his animal faces, than in using animal characteristics to explore human society".[21] Using animals as ciphers, or "place-fillers", to refer indirectly to humans and allegorise human concerns, tallies with a perceived anthropocentric bias in the representations and readings of anthropomorphised animals in comic books.[22]

When animal characters are heavily humanised, minimising the animal presence to the point of theriocephaly or reverse anthropomorphism, as they are in the funny animals genre, they can encourage readers to overlook or see through the animal. In Brian Cremins' survey of funny animals, he argues that "animal tales and fables are not about real animals at all" and holds that "the strange creatures, for all their comic appeal, also remind us of how it feels to be fragile, vulnerable, and human".[23] It is apparent that anthropomorphic images are not predominantly about engaging with other animals, but rather to give readers pause to reflect on what it means to be human. Michael Chaney furthers this anthropocentric interpretation in his densely theoretical and rewarding essay, "Animal Subjects of the Graphic Novel" (2011), in which he argues that not only is the human usually imbibed in non-human animal exteriors that always bespeak the human, the superiority of the human is also celebrated and naturalised in its effacement. The animal "functions as mask or costume", he argues, "beneath which lies the human, whose universality is reaffirmed and reified in the process".[24] The anthropomorphic, or theriocephalic image, initiates a process of detecting the human through transformation, despite the animal veil: "Rather than reading the human body as it is, we are often called upon in comics to view and imagine it as dramatically other than it is. But this reading protocol of inversion is precisely also the origin of the radically conservative politics behind visual distortion".[25] It follows in theory that the more images amalgamate the human with the animal, the more readers are compelled to restore the obscured human. So, some paradigm of the human is the focus of graphic narratives even when it does not appear to be the case on face value.

The conservative humanist reading of animal images in graphic narratives is applicable in part to *Blacksad* and *Grandville*. The long exposure to funny animal characters, from Krazy Kat (1913) and Tiger Tim (1919) to *Kung Fu Panda* (2008) and *Zootopia* (2016), means readers are habituated to assimilating animal heads into anthropocentric approaches. There is no assumption that these are supposed to depict real animals; there is little experience of disbelief. In *Blacksad* and *Grandville*, the animal faces appear as token identities frequently, as the majority of the stylistic and narrative references construct humanised characters and concerns, which can overwhelm the animal. The anthropocentricism embedded in this imbalance is prompted by the environmental, discursive and generic contexts that the anthropomorphised characters engage in, which are all oriented towards a familiar human world, making it difficult for the animal heads to register as aberrations for lengthy periods. As readers become accustomed to looking through the animal image to access the human, the implication is that the animal eventually vanishes from the conscious reading experience.

This immersion in a predominantly human world is complicated in numerous ways in *Blacksad* and particularly in *Grandville*. Canales and Talbot incorporate references to the animality of their protagonists to challenge an undisturbed reading of the human. These references are regularly in the name of humour and rely on the reader bearing the animality of the character in mind to function properly. Early in "Somewhere within the Shadows", for example, there is irony in the alabaster statue of a black cat in Blacksad's office, while in *Grandville*, there is a similarly ironic statue of the Egyptian cat goddess, Bastet, in the foreground of panels.[26] In these worlds of theriocephalic characters, such statues would stand as naturalistic representations, which is why they are unremarkable to Blacksad and LeBrock. For readers, however, the statues allude to the preoccupation in human cultures with surrounding ourselves with artificial animals. In his essay, "Why Look at Animals" (1980), John Berger describes this phenomenon as a compensatory gesture driven by some deep nostalgia to fill the void left by the ontological separation of man and animal.[27] As such, the statues serve as a reminder of Blacksad and LeBrock's connection to a historical tradition of anthropomorphism and their construction as hybrid animal–humans.

Blacksad also recognises himself as a "real" cat in both explicit and implicit ways within the fictive reality of the story world. He acknowledges that "There are a lot of clichés about us cats. 9 lives. Never wanted to find out if that's true" (43). In "Red Soul", when Alma (who resembles catwoman from the Batman franchise more than an anthropomorphised cat) points out that there is a superstitious side to Blacksad, he replies "what else can an old black cat like me believe in" (161). Elsewhere, he also puns on his species, saying "I wasn't some defenceless pussy", which doubles up as the vernacular one might expect in a contemporary noir detective story (25). Sometimes the jokes depend on the common understanding of a cat's relationship with other animals, such as when Blacksad confesses

"Something personal against rats. That's true" or when his sidekick Weekly the weasel is at once alarmed and confused when Dinah calls Blacksad a "son of a bitch", which, taken literally, is a peculiar parentage for a cat and would make Blacksad related to the same Caniformia suborder as Weekly (43, 76). The fact returns in these moments that Blacksad identifies as a cat and is identified by others as a cat, which revises his animal status from a nominal or visual façade to a constituent part of his characterisation. Despite the fact that these references draw upon human observations and perceptions of cats, they work to reinstate the animal's presence at the forefront of the narrative briefly to go against the humanist grain supported by the setting, language and storylines.

Whereas *Blacksad* is punctuated with reminders of the animal, such references are less frequent in *Grandville*, although many of the characters' names incorporate words for animals, including LeBrock and Sarah Blaireau, which are the dialectal and French terms for "badger". While these names suggest the characters are what they are called, it is telling that Talbot declares wryly in the credits that "No animals were harmed in the making of this book", as it alerts the reader to the gulf between animal images and real animals (10). Talbot is implying here that the extreme violence depicted in his book does not directly impact animals (unlike the harmful practices to real animals in other mediums, mainly film), even if he does naively overlook any ideological implications his depictions might have, particularly in continuing to associate non-human animals with violence. After this declaration, it is perhaps unsurprising that self-reflexive animal references are mainly limited to a few general idioms, such as "sent him away with a flea in his ear" or "frightened sheep ready to follow their shepherd" (40, 61). This relative dearth is explained somewhat by the fact that badgers have fewer associated tropes and proverbs on which to draw compared to cats.

It is the inclusion of human beings in *Grandville* that establishes the anthropomorphic characters as at least partial animals. When LeBrock and his adjunct Roland Ratzi first arrive in Grandville to investigate the murder of Raymond Leigh-Otter, it is made known that a reversal of the status of human and non-human animals is at work in the book, as humans are classed as citizens without rights and perceived as curious others. They are "menial workers" that toil alongside servant robots and are referred to as "doughfaces" (19). The disparaging word "doughface" suggests a malleable identity, which is related to how the term "doughfacism" was employed in America in the mid-nineteenth century, when it described "the willingness to be led about by one of stronger mind and will" according to the 1847 Webster's dictionary. The inverted human and non-human animal hierarchy, reminiscent of *The Planet of the Apes* premise, is confirmed by the acknowledgement that the humans are distant relatives of the animals, as a "hairless breed of chimpanzee", which means the subjugation of humans is in spite of a known evolutionary continuum between human and non-human animals, thereby mirroring an anthropocentric reality (19). This alternative, transposed perspective evokes Michel de Montaigne's famous insight in his "An Apology for Raymond Sebond" where he observes that "we understand them no more than they us. By the same reason, may they as well esteeme us beasts as we them".[28] The implicit equation here is that radical difference between species results in incomprehension and this in turn causes devaluation. The peculiar upshot is that human and non-human animals are rendered similar at the very point of their difference, owing to their shared egocentrism and subordination of others. In effect, then, Talbot's allegorical strategy of redirecting speciesist discriminations from non-human animals to humans ultimately exposes the long-standing immoral treatment of animals in two steps: eliciting the disturbing possibility of humans being the subaltern subjects, in conjunction with the logic that the humans represent non-human animals and the animal heads represent humans in the topsy-turvy world of *Grandville*.

Hermeneutic Animals: Human Logos in Graphic Detective Fiction

Understood as humans with animal heads and sometimes animals with human bodies, *Blacksad* and *Grandville* show that their detectives are of indeterminate ontological status. Owing to this kind of unsettled identity, Glenn Wilmott writes: "the funny animal may be regarded as something of a biomorphic question mark, which elsewhere I have called a 'problem creature'".[29] The detective fiction genre, however, lends weight to the idea of the animal image as a translucent disguise, through which the human being is visible. The detective's most basic role, originating from the word "detect", is to discern and discover. Detective fiction generally has a strong focus on human agents utilising empirical observation, memory, and deductive and inductive reasoning to solve complex cases. The *modus operandi* to achieve success is highly variable, and can include dogged persistence, underhand tactics, intimidation and outright violence, but it is reasonable to generalise that cohering information and accruing knowledge is central to the detective fiction genre. Although ethnologists and ethicists such as Frans de Waal and Peter Singer make compelling arguments for the rationality and self-consciousness of some non-human animals to the point of personhood and moral rights, it is clear that neither would extend these capabilities to the level required for typical detective work.[30]

This is not to say that real animals and human perceptions of animals are incongruous to other aspects of detective fiction. As discussed above, there are several actual and apocryphal attributes that make some animals more suitable candidates for anthropomorphised detectives and, through the lens of critical anthropomorphism, might be said to identify shared or related properties between human and non-human animals. In terms of a more traditional anthropocentric schema, it would also be fitting to have the degenerate, criminal underworld of urban societies populated by animalised figures to convey economically a wild, unreasoned, immoral strata; something like the Foucauldian *gyrllos*, emblematising the animal madness secreted in the recesses of the human mind.[31] As Blacksad informs us, this is "my world. A jungle where it's survival of the fittest. Where people act like animals" (56). Animality is an established metaphor for manifestations of these primal, savage aspects of humanity.

Blacksad and LeBrock are not exempt from this feral world either. Firstly, they are not strictly classic detectives from the Golden Age or later cosy mysteries. Blacksad derives from the tough, cynical, hard-boiled mould and LeBrock has both brains and brawn as a Holmes type mixed with a fierce powerhouse, as evident in the panel of him reading about the French private detective Vidocq while simultaneously weightlifting (45). This means that, in some respects, both detectives are in close proximity with the violence and corruption they oppose. Blacksad's exposure to crime and conspiracy, what he calls "evil", causes him to assert: "I had chosen to walk the dark path in life and I'm still on it" (63, 56). He readily acknowledges that his own virtue is questionable when reflecting on the "civilised person I used to be" and wondering "who can call themselves moral these days" (13, 48). Similar to Blacksad's uncomfortable acquaintance with moral regression, LeBrock is a brutal enforcer and breaker of the law, a contradiction that finds its greatest expression when he binds the corrupt Archbishop of Paris (a chimpanzee), threatens to burn him in a pool of brandy and slices off his ear (58–63). This scene is an allusion to

Tarantino's *Reservoir Dogs* (1992) in which the psychopathic Vic Vega cuts off a kidnapped LAPD police officer's ear and douses him in petrol. The reference pairs LeBrock with the deranged criminal to highlight the thin line between right and wrong, order and disorder. For Blacksad and LeBrock, the animal can symbolise their uncultivated methods and traversal of a slippery moral tightrope.

While the menagerie certainly holds for the darker sides of the characters and the atavistic elements of society, it is not necessarily appropriate for the more ameliorative, redemptive vision of the detective as an astute problem-solver. To differing extents, both detectives depart from the early paradigm that Hammett describes in his introduction to *The Maltese Falcon* as "the erudite solver of riddles in the Sherlock Holmes manner".[32] Yet Blacksad and LeBrock do bear the legacy of classical detective fiction in the way they exhibit exceptional mental acuity to progress their investigations. They are knowledgeable and perceptive, streetwise and shrewd, as well as being world-weary and wisecracking. In "Arctic Nation", for instance, Blacksad observes that snow on a staircase has been swept to cover footprints, which alerts him to danger (84). In "Red Soul", he openly states that the "best quality in a good detective isn't that they're a crack shot or that they're in great shape, but that they're a quick thinker" and he articulates his role as an assembler of intelligibility: "I needed something else. A piece of the puzzle that would help me see the whole picture" (170, 142). Likewise, LeBrock employs close attention and logical thinking to spot clues and solve mysteries that escape all others, such as when he recognises a combination of clean and dusty surfaces as signs of concealment in Leigh-Otter's apparent suicide (13–15). More impressively, he identifies an Agapanthus flower in full bloom in the background of a photograph and is consequently able to determine the timeframe as late July/August (51).

As the examples of ingenious feats of observation, memory recall and deduction above attest, the image of the animal is inherently incongruous to an abiding element of detective fiction to the extent that animal characters can serve as a generic parody. When the head is understood as the seat of the mind, having an animal head emphasises, even pokes fun at, the extent to which classical detective fiction in particular depends on "ratiocination" to reconstruct past events and formulate a solution to the mystery. Lee Horsley identifies "two central elements of Holmes's method" as "a scientific approach rooted in a Victorian faith in the accumulation and cataloguing of data, and rational and logical analysis based on this scientific foundation".[33] In this Holmesian model, the prior accrual of vast amounts of information and the ability to read indexical signs live at a crime scene result in a mental picture of the circumstances as they transpired, a lead on the culprit and an (eventual) revelation.

One of the appeals of detective fiction is reader participation, especially in narratives that comply with Knox's rules of fair play, as the receiver of the text, much like the detective, is invited to combine past knowledge of the genre and the interpretation of inscribed signs to assess potential scenarios and culprits.[34] Carl D. Malmgren notes in *Anatomy of Murder* how detective fiction reveals "a subconscious desire to treat the world as if it were a book", exercising the hermeneutic animal in the act of *reading*: to study, decipher and foresee.[35] Approaching the world as textual also complements the reflective and biographical inclination of humans, as Cremins points out: "The animal within us is an obsessive one, a being who studies the past in order to make some sense of the present".[36] In conjunction, methods of detection and their parity with reading imply an appreciation of humanist principles, as they depend on the reason of the individual to recuperate order, both in terms of trying to solve a problem to fill the gap in knowledge and, in most cases, apprehending the culprit in an attempt to correct the moral imbalance of the crime. In direct contrast to the human as hermeneutic and retrospective, non-human animals have been perceived as tethered to the immediate now. Akira Mizuta Lippit evokes the Nietzschean amnesiac animal in observing that, "by forgetting to speak, erasing history, the animal lives happily, in the present and without secrets".[37] Even if violence is second nature to the detective and pervasive corruption makes it seem futile to endeavour to restore order, in terms of the thoughtful and reflective aspects of detection, an animal is not optimal and seems to play with the human preoccupation with enlightenment.

On balance, an animalised detective is at odds with the intellectual principles of detection that rest on problem-solving logic and critical thinking. These core conventions of "detecting" would seem risible if readers were to think of real animals in these roles, because, even in the most basic investigations, they involve complex forms of thinking (abstract, recursive, divergent, convergent, sequential and holistic), which are different in degree for humans, if not always in kind. This conflict raises a dichotomy between the surface appearance of the animal and the complex depth of the human mind, whereby the caricatured animal head becomes a superficial theriocephalic mask covering human reasoning capacities. Notwithstanding the animal references that punctuate both *Blacksad* and

Grandville to varying degrees, the detective fiction genre encourages the reader to look through the animal head for the human cognition within. In other words, the animal is treated as mere *image* and the human as the substantive *logos*, a term which includes logic and reason as principles of knowledge, judgement and order among its many related meanings from Heraclitean philosophy to Jungian psychology. In effect, then, *Blacksad* and *Grandville* show that the mimetic representation of the animal combines with the animal's role as a disguise thinly veiling the human to suggest real animals are twice removed in graphic detective fiction.

When framed as "animal as surface image and human as deep logos", a reasonable analogue is the philosophical theory of Cartesian dualism, as it recognises the body and mind as two distinct entities associated with animal and human respectively. René Descartes' discredited but still profoundly influential systematic account proposes that "there are two different principles causing our motions: one is purely mechanical and corporeal and depends solely on the force of the spirits and the construction of our organs, and can be called the corporeal soul: the other is the incorporeal mind, the soul which I have defined as a thinking substance".[38] He continues to assert that the presence of a thinking soul cannot be proved in the animal, and that animals might be deemed "natural automata", or organic machines.[39] In this Cartesian formulation, the animal is the mechanical and corporeal, or outer body, whereas the human is the incorporeal thinking substance, or inner mind. The anthropomorphic tradition that attributes predominantly human characteristics to animalistic forms, and the theriocephalic mask that superimposes an animalistic form over predominantly human characteristics, both indicate an appropriation of the animal as surface image/outer body and human as deep logos/inner mind. As comic books, *Blacksad* and *Grandville* allude to this dualism to a further extent through their formal components in the distinction between image and text, drawing on the linguistic, discursive element embedded in the term "logos", in contradistinction to the visual element of graphic narratives. That is, the human is the written language, read and understood, discrete from the animal as an image, observed and perceived.

The theriocephalic image in graphic detective fiction forges a series of linked binary opposites from animal/human to surface/depth to body/mind to image/text. However, closely considering the complex interplay between texts and images in comic books presents a more nuanced relationship between the two, in which the graphic is "not an illustration of a script, but the making of an autonomous story world built by the creative interaction of words and images".[40] A revised understanding of these formal elements means the comic book medium itself can intimate a more dynamic, multifarious model that can ripple back through the chain of binaries, effectively working in reverse to encompass the various and coexisting relationships between human and non-human animals.

In his introduction to *The Future of Text and Image*, W. J. T. Mitchell draws on his previous works including *Iconology: Image, Text, Ideology* (1986) and *Picture Theory* (1994) to describe the variable interactions between texts and images. He outlines the view that there are "normal and normative relations between texts and images. One illustrates or explains or names or describes or ornaments the other. They complement and supplement one another, simultaneously completing and extending".[41] And yet Mitchell argues that texts and images actually have three potential relations, symbolised by a slash, a hyphen and a compound word: "the image/text rupture, the image-text relation, and the imagetext synthesis".[42] When understood as rupture, relation and synthesis, three levels of difference and distance concerning images and texts emerge. They variously or simultaneously operate separately, dialogically or in union. The "image/text rupture" recognises something of the crude opposition of texts and images, whereas the "imagetext synthesis" is possible for Mitchell as pictures can be "textual" and texts can be "visual"; there is nothing fundamentally different about the ideology of the two, or how we speak about them, on the level of semiotics.

The rupture, relation and synthesis of images and texts is significant in the context of animals in comic books owing to the revised binary logic at work, particularly as it partially deconstructs the animal as image/body/surface and human as logos/mind/depth formulation in graphic detective fiction such as *Blacksad* and *Grandville*. Chaney makes the apposite point that "comics routinely problematize the human by blurring the ontological boundary between humans and animals according to the same logic that both fuses and separates words and pictures".[43] If image and text are not purely characterised by rupture, but also by relation and synthesis, then the comic book form indicates that it exceeds strict dualism to explore the connected and shared ground between textual and visual mediums.

A potent example of these various formal relationships is the use of onomatopoeia to bring visual action to life through text. In one fight scene in "Red Soul", the words "Splat!" and "Splotch!" accompany Blacksad punching and drenching three gang members (167). In a nod to this outdated comic book convention, Blacksad remarks, "My bag of tricks was evidently in need of a restock" (167). Having text appear outside of the speech bubbles or voiceover captions presents an overt example of formal fusion and separation at once. The illustrations of Blacksad swinging and throwing indicate movement and a resulting sound; the words represent the types of sound created; and the images and text combine to give a more visceral sense of action. This is an accepted immersive strategy, yet the text is also part of the visual composition and can seem out of place, both as an outdated convention when applied to the realism of a hard-boiled detective story and if read literally as existing in the rendered space as text, not sound. In this way, comic books can give the impression that image and text are at times consistent and conflicting.

Like the varied relations between the comic book's formal components, in which images and texts can work independently, co-dependently and as one, human and non-human animals are also different, similar and the same. They have distinctions, equivalents and identical aspects. The formal potential for rupture, relation and synthesis therefore helps to variegate the interactions within the texts' content too, to propound a structure of kinship and identification between the human and animal presences, in addition to disparity. This means that as texts and images become metonymic for the human and animal, the animal's reduction to surface appearance, particularly in contrast to detective fiction's emphasis on the human as a reasoning creature, is counterbalanced by the relationship between words and visuals, which suggest a complementary and necessary presence in the way they generate meaning together. It is this deep formal symbiosis in graphic narratives that projects a continuum between human and non-human animals to mitigate the arguably anthropocentric ideological structure of "animal as cipher for the human" in the funny animals genre. In this context, Blacksad's utterance, "I've endured all kinds of humiliation but I always thought about the day when I would get even", resounds with an anti-speciesist sentiment in which the rendering of hybrid creatures is not recognised indiscriminately as anthropomorphic, fixated primarily on exclusively human characteristics, but open to a more equitable perception of the variously idiosyncratic, interpenetrating and indistinguishable characteristics of living creatures (43).

Blacksad and *Grandville* are examples of a nominal kind of visual anthropomorphism that present barely animalised humans, while also precluding a completely anthropocentric reading perspective. As graphic detective fictions, they further exhibit the opposition of animal image and human logos through the prioritisation of inner human "ratiocination" in contradistinction to the outer theriocephalic mask. Following this division, *Blacksad* and *Grandville* inspire a chain of related binaries, the end point of which is the formal composition of the comics themselves, and yet the various potential relationships between images and words write back against any simple oppositional structure to redress the multiplicity of human–animal relations and restore ambivalence to the anthropomorphic tradition.

Notes

1. Wilson, "Who Cares?" 39; Doherty, "Art Spiegelman's Maus", 71; Baker, *Picturing the Beast*, 174.
2. Gelly, "Doyle's Sherlock Holmes Stories".
3. There are numerous other examples of animals as investigators or appearing in graphic detective fiction, including Robert Crumb's Fritz the Cat in "Secret Agent for the C.I.A." (1965); Tom Stazer's feline investigative journalist Lionheart in *Critters* (1986); the hippopotamus–human hybrid, Hieronymus Flask, in Comicraft advertisements in the 1990s and later the dystopian science fiction series *Hip Flask* (2002); the animals investigating paranormal activity in Evan Dorkin and Jill Thompson's *Beasts of Burden* (2003–2016); and Howard the Duck, who sets up as a private eye in 2015.
4. Rifas, "Funny Animal Comics", 234.
5. Steranko, "Introduction", 7.
6. Mitchell, "Introduction", 9.
7. Talbot, "Grandville and the Anthropomorphic Tradition".
8. Steranko, "Introduction", 7.
9. Talbot, *Grandville*,10.
10. See Talbot, *Grandville*, 25–6 for LeBrock's violence; see 39 and 95 for his demonic appearance.
11. Holme, "Introduction", 5.
12. Hawthorne, "Bryan Talbot's *Grandville*", 49.
13. Steranko, "Introduction", 7.
14. Talbot, "Grandville and the Anthropomorphic Tradition".
15. Talbot, "Grandville: the annotations".
16. Heath Justice, *Badger*, 104.
17. Unless otherwise stated, all definitions taken from *OED Online* (Oxford University Press, March 2018). For more on badger baiting and melicide, see Justice, *Badger*, 146.
18. Keen, "Fast Tracks to Narrative Empathy", 138.
19. Brown, "The Speaking Animal", 74.

20. In particular, see panel 5 in Canales and Guarnido, *Blacksad*, 15 and panel 1 in Talbot, *Grandville*, 48.
21. Hawthorne, "Bryan Talbot's *Grandville*" 132.
22. Tyler, *CIFERAE*, 4.
23. Cremins, "Funny Animals", 148; 146.
24. Chaney, "Animal Subjects", 132.
25. Chaney, "Animal Subjects" 132.
26. Canales and Guarnido, *Blacksad*, 13; Talbot, *Grandville*, 45, 67.
27. See Berger, "Why Look At Animals?", 251–261.
28. de Montaigne, "An Apology for Raymond Sebond", 58.
29. Willmott, "The Animalised Character and Style", 54.
30. For more on animal intelligence and moral status, see Frans de Waal, *Are We Smart Enough?*; Singer, *Practical Ethics.*
31. Michel Foucault refers to the image of the *gryllos*, a figure with two or more faces (sometimes a combination of young and old men, satyrs and animals) as part of his discussion on madness, unreason and animality in *Madness and Civilization*. See Foucault, *Madness and Civilization*, 17–19.
32. Hammett, "Introduction", viii.
33. Horsley, *Twentieth-Century Crime Fiction*, 40.
34. Knox, "Preface", 12–16.
35. Malmgren, *Anatomy of Murder*, 47.
36. Cremins, "Funny Animals", 152.
37. Lippit, "Magnetic Animal", 1116. For extracts and a concise discussion of animals in Friedrich Nietzsche's philosophy, see Nietzsche, "O My Animals", 1–14.
38. Descartes, "Animals are Machines", 284–5.
39. Descartes, "Animals are Machines", 285.
40. Baetens and Frey, *The Graphic Novel*, 50.
41. Mitchell, "Introduction: Image X Text", 5.
42. Mitchell, "Introduction: Image X Text", 9.
43. Chaney, "Animal Subjects", 133.

Works Cited

Baetens, Jan and Frey, Hugo. *The Graphic Novel: An Introduction.* Cambridge: Cambridge University Press, 2015.

Baker, Steve. *Picturing the Beast: Animals, Identity, and Representation.* Manchester: Manchester University Press, 1993.

Berger, John. "Why Look At Animals?" In *The Animals Reader*, edited by Linda Kalof and Amy Fitzgerald, 251–261. New York: Berg, 2007.

Brown, Lisa. "The Speaking Animal: Non-human Voices in Comics," In *Speaking for Animals: Animal Autobiographical Writing*, edited by Margo DeMello, 73–78. Oxford: Routledge, 2013.

Canales, Juan Diaz and Guarnido, Juanjo. *Blacksad*. Milwaukie: Dark Horse, 2010.

Chaney, Michael A. "Animal Subjects of the Graphic Novel." *College Literature* 38, no. 3 (2011): 131–149.

Cremins, Brian. "Funny Animals." In *The Routledge Companion to Comics*, edited by Frank Bramlett, Roy T Cook and Aaron Meskin, 146–153. New York and London: Routledge, 2017.

Descartes, René. "Animals are Machines." In *Environmental Ethics: Divergence and Convergence*, edited by S. J. Armstrong and R. G. Botzler, 281-285. New York: McGraw-Hill, 1993.

Doherty, Thomas. "Art Spiegelman's Maus: Graphic Art and the Holocaust." *American Literature*, 68, no 1 (1996): 69–84.

Foucault, Michel. *Madness and Civilization*. London: Routledge, 2001.

Gelly, Christophe. "Sir Arthur Conan Doyle's Sherlock Holmes Stories: Crime and Mystery from the Text to the Illustrations." *Cahiers victoriens et édouardiens*, Vol 73 (2011): 93–106.

Hammett, Dashiel. "Introduction." In *Maltese Falcon*. New York: Modern Library, 1934.

Hawthorne, Melanie. "Bryan Talbot's Grandville and French Steampunk." *Contemporary French Civilization* 38, no 1 (2013): 47–72.

Holme, Bryan. "Introduction." In *Grandville's Animals: The Word's Vaudeville*. New York and London: Thames and Hudson, 1981.

Horsley, Lee. *Twentieth-Century Crime Fiction*. Oxford: Oxford University Press, 2005.

Justice, Daniel Heath. *Badger*. London: Reaktion, 2015.

Keen, Suzanne. "Fast Tracks to Narrative Empathy: Anthropomorphism and Dehumanization in Graphic Narratives." *SubStance*, 40, no 1 (2011): 135–155.

Knox, Ronald. "Preface." In *The Best Detective Stories of 1928-29*, edited by Ronald Knox, 9–23. New York: Horace Liveright, 1928.

Lippit, Akira Mizuta. "Magnetic Animal: Derrida, Wildlife, Animetaphor." *MLN*, 113, no 5 (1998): 1111–1125.

Malmgren, Carl D. *Anatomy of Murder: Mystery, Detective and Crime Fiction*. Madison: University of Wisconsin, 2001.

Mitchell, W. J. T. "Introduction: Image X Text." In *The Future of Text and Image: Collected Essays on Literary and Visual Conjunctures,* edited by Ofra Amihay and Lauren Walsh, 1–14. Newcastle: Cambridge Scholars, 2012.

Montaigne, Michel De. "An Apology for Raymond Sebond." In *The Animals Reader*, edited by Linda Kalof and Amy Fitzgerald. 57–58. Oxford and New York: Berg, 2007.

Nietzsche, Friedrich. "O My Animals." In *Animal Philosophy*, edited by Peter Atterton and Matthew Calarco, 3-6. London: Continuum, 2004.

Rifas, Leonard. "Funny Animal Comics." In *Encyclopedia of Comic Books and Graphic Novels* 1, edited by M. Keith Booker, 234–242. Oxford: Greenwood, 2010.

Scaggs, John. *Crime Fiction*. London and New York: Routledge, 2005.

Singer, Peter. *Practical Ethics*. Cambridge: Cambridge University Press, 2011.

Steranko, Jim. "Introduction." In *Blacksad*, by Juan Diaz Canales and Juanjo Guarnido. Milwaukie: Dark Horse, 2010.

Talbot, Bryan. *Grandville*. London: Jonathan Cape, 2009.

Talbot, Bryan. "Grandville and the Anthropomorphic Tradition." Northumbria University 20th Anniversary Public Lecture Series, January 21, 2013.

Talbot, Bryan. "Grandville: the annotations." *The Official Bryan Talbot Website*, accessed 1 March 2018. https://www.bryan-talbot.com/grandvilledirectors-cut/index.php

Tyler, Tom. *CIFERAE: A Bestiary in Five Fingers*. Minneapolis: Minnesota University Press, 2012.

Waal, Frans de. *Are We Smart Enough to Know How Smart Animals Are?* London and New York: Norton, 2016.

Willmott, Glenn. "The Animalised Character and Style." In *Animal Comics: Multispecies Storyworlds in Graphic Narratives*, edited by David Herman, 53–76. London: Bloomsbury, 2017.

Wilson, Edmund. "Who Cares Who Killed Roger Ackroyd?" *New Yorker*, January 20, 1945, 39–40.

"As easy to spot as a kangaroo in a dinner jacket": Animetaphor in Raymond Chandler and Jonathan Lethem

Ruth Hawthorn

In developing the hard-boiled style through his iconic Philip Marlowe series, Raymond Chandler hoped to redeem detective fiction from being "a cheap, shoddy and utterly lost kind of writing" and move it into the realms of "something that intellectuals claw each other about".[1] Disparaging plot-focused narratives, he wrote that "the formula doesn't matter, the thing that counts is what you do with the formula; that is to say, it is all a matter of style".[2] As Stephen Tanner has observed, Chandler's similes—" comically incongruous, but at the same time vividly accurate"—are a hallmark of that style and an important part of Chandler's mission to capture a democratic American vernacular. [3] David Smith argues that this language is integral to the character's anti-authoritarian stance and his grim cynicism:

R. Hawthorn (✉)
University of Lincoln, Lincoln, UK
e-mail: rhawthorn@lincoln.ac.uk

R. Hawthorn, J. Miller (eds.), *Animals in Detective Fiction*, Palgrave Studies in Animals and Literature,
https://doi.org/10.1007/978-3-031-09241-1_14

> So long as Marlowe can tell us that he can listen to "spring rustling in the air, like a paper bag blowing along a concrete sidewalk" then we know his mind can cope with, and see through, any inflated boosterism that the world claims as reality.[4]

The ostentatious wordplay, then, has a disruptive function; to be more geographically specific, it highlights the dark (sur)reality of Los Angeles, a city which Michael Sorkin describes as "the most mediated town in America".[5] Arguing along similar lines, Mike Davis writes: "To move to Lotusland is to sever connection with national reality, to lose historical and experiential footing, to surrender critical distance, and to submerge oneself in spectacle and fraud".[6] Chandler's work does not just describe but rather creates Los Angeles, contributing to the noir counter-mythology of a city which—as many others have argued—had come to symbolise the climax of several, inter-related, triumphant nationalist narratives: westward expansion, American Exceptionalism, Manifest Destiny and the Dream of self-sufficiency.[7]

In particular, animal imagery is integral to Chandler's disruptive style. From the opening of his first novel, *The Big Sleep* (1939), where Marlowe encounters "trees trimmed as carefully as poodle dogs", to his predatory femme fatales, like Velma Valentino with her "darting snake" tongue, animal figuration allows Chandler to skewer his subject with linguistic economy and wit, and, more importantly here, to unsettle the human/non-human animal hierarchy.[8] Akira Mizuta Lippit, who has written extensively on the figurative use of animals, argues that, linguistically, animals present a peculiar and specific phenomenon. He finds a

> fantastic transversality at work between the animal and the metaphor—the animal is already a metaphor, the metaphor an animal. Together they transport to language, breathe into language, the vitality of another life, another expression: animal and metaphor, a metaphor made flesh, a living metaphor that is by definition not a metaphor, antimetaphor—"animetaphor".[9]

Lippit argues (following Nietzsche, Agamben and Derrida) that animals, traditionally denied subjectivity in Western philosophy in large part because they are thought to lack language, point to the very limits of language: "On the verge of words, the animal emits instead a stream of cries, affects, spirits, and magnetic fluids. What flows from the animal touches language without entering it, dissolving memory, like the unconscious into a timeless present".[10] Animals, here, remain forever in the present moment, unreachable through the human language which constructs memory and would place them in a narrative continuum. In Lippit's view, animals' proximity to language, without ever themselves commanding it, makes them ultimately unassimilable; as Eva Hayward puts it, "animals are always troubling the language that attempts to name them".[11] In Chandler's case, these animetaphors work to unsettle the reverence around the category of "human", showing up characters' absurdity or monstrosity and highlighting Chandler's existential pessimism.

In Jonathan Lethem's *Gun, with Occasional Music* (1994), however, Chandler's metaphorical, non-present animals are actualised. Taking for his epigraph one of Chandler's characteristic wisecracks from the late novel *Playback* (1958)—" the subject was as easy to spot as a kangaroo in a dinner jacket"—the dystopian society of Lethem's hybrid noir-sci-fi is home to "evolved" animals who have been genetically developed to act like humans, but who are treated like second-class citizens, including the detective's nemesis, the kangaroo-gangster, Joey Castle.[12] Looking at Lethem's appreciative pastiche of Chandler's hard-boiled style, which allows him to build his dystopian world with sharp and disorienting brevity, this chapter explores the disruptive function of Lethem's literal animals in the novel's aggressively policed present, where citizens are kept in a state of government-sanctioned amnesia. Drawing together these strands on animetaphor and memory, I argue that the novel unsettles the assumption that humans constitute an evolutionary pinnacle, supporting Deborah Bird Rose's contention that noir—with its focus on self-destructive protagonists and its blurring of the lines between criminal and victim—is an exemplary genre for criticising the problems of the Anthropocene where "our Western eagerness to remake everything in the name of progress and emancipation is bringing us ever closer to ruin".[13]

Chandler's Noir Animals

In order to understand Lethem's pastiche, we first need to look in closer detail at how Chandler's animal metaphors work.[14] I suggest that they fall into two broad categories which serve two slightly different, but ultimately related, functions. The first of these categories is animal imagery which is used to create a sense of the surreal. As Effie Rentzou argues, Surrealism fundamentally questions "human singularity, agency, and even the human unconscious by drawing continuous lines between man and animal".[15]

The subject being "as easy to spot as a kangaroo in a dinner jacket" is an obvious example of this kind of boundary dissolution and the opening of the earlier novel *Farewell, My Lovely* (1940) offers a very similar coinage. Marlowe is drawn into the case through his fascinated observation of a complete stranger who is described as follows: "Even on Central Avenue, not the quietest dressed street in the world, he looked about as inconspicuous as a tarantula on a slice of angel food."[16] And again, later in the same novel, Marlowe sums up his own mental state as being "crazy as two waltzing mice".[17] In each case, the animal serves to highlight the incongruous, the idiosyncratic. The kangaroo, the tarantula, the mice are all compelled into situations where they are utterly out of place. This forced yoking together of the non-human animal with frivolous human constructs (a sartorially elegant marsupial, the terrifying arachnid on a fluffy cake, rodents observing dance steps) creates a defamiliarising humour, succinctly texturing Chandler's disorientating fictional world. More particularly, the unsettling presence of the animals in these images is used to suggest the absurd triviality of the human endeavour in each example, contributing to the misanthropy of Chandler's noir world view to which I will return shortly.

Contrary to the Surrealist-inflected passages above, in the essay "The Simple Art of Murder", which has become an unofficial manifesto for the hard-boiled style, Chandler actually claimed a greater commitment to realism for American detective fiction. He was scathing of what he saw as the overworn English vein of mystery story, populated with characters who "must very soon do unreal things in order to form the artificial pattern required by the plot, [ceasing] to be real themselves".[18] In opposition to this, he championed Dashiell Hammett who, he argued:

> gave murder back to the kind of people that commit it for reasons, not just to provide a corpse; and with the means at hand, not with hand-wrought duelling pistols, curare, and tropical fish. He put these people down on paper as they are, and he made them talk and think in the language they customarily used for these purposes.[19]

This claim of authenticity had been used as a promotional tool by pulp editors like Joseph T. Shaw long before Chandler's essay but this realism is itself, of course, a carefully crafted literary effect and hard-boiled fiction soon established its own set of readily recyclable generic tropes.[20] As Sean McCann observes:

> Like the secluded mansions of the golden-age detective story, their [hard-boiled writers'] urban landscape was a dreamworld, an exaggerated image of the metropolis as a battlefield of crime lords and corrupt officials that drew as much from the traditions of pop imagery and urban folklore as it did from direct observation.[21]

The kinds of incongruous animal figurations I have been discussing so far are a key part of this "exaggerated image", this dreamworld quality McCann attributes to Chandler's work. They stand out as garish and absurd, they draw attention to their own literariness and are really quite far from the language people "talk and think in"; rather, they are part of what Lethem admires as Chandler's "vernacular surrealism".[22]

The second category I want to look at is the use of extended animal metaphor to build character. The most obvious examples of this are Chandler's femme fatales. Chandler is not a writer who is known for his feminist credentials and his murderous women are garishly rendered, drawing on well-worn reptilian and fiendish motifs. Carmen Sternwood's deviance, for instance, is mapped onto imagery of cats and snakes. Her father claims that—like the rest of her family—she has "no more moral sense than a cat"; she is described as having "little sharp predatory teeth" and in the notorious scene where Marlowe rejects her advances, a "*hissing noise came tearing out of her mouth as if she had nothing to do with it*".[23] This hissing returns at the denouement where the point is made even more explicit: "The hissing sound grew louder and her face had the scraped bone look. Aged, deteriorated, become animal, and not a nice animal".[24] Carmen's villainy, then, is imagined in terms of animality; the animalistic tropes become a shorthand or cipher for her amorality, tapping into a much longer tradition of discourse linking criminality and animality. As Greta Olson argues, such language "reifies species divisions: normative humans do not partake in 'beastly' crimes; animals and animalistic humans, by contrast, do".[25] Cats and snakes are creatures commonly associated with duplicity, selfishness, original sin and their deployment here is an example of what John Simons describes as literature "appropriating the non-human experience as an index of humanness".[26] The animals are not there *as animals* but are used for what they can tell us about human behaviours. More specifically, the use of animality speaks to Marlowe's (and Chandler's) misogyny; it is an important part of his rhetorical arsenal for othering women, which is also grimly apparent in what Megan Abbott has called his "taxonomy of blondes" in *The Long Goodbye* (1953), where women are sneeringly reduced to types, such as "the small cute blonde, who cheeps and twitters" or the "big statuesque blonde who straight-arms you with an ice-blue glare".[27] The specimens of blonde are subjected to a mock-scientific evaluation for the enticements and risks they present the homosocially interpolated "you", centring the implied straight, white, male reader against this menagerie of potentially dangerous females.[28]

What is central to both categories of animal figuration in Chandler's texts is that they are used to unsettle the anthropocentric assumption of human preeminence. The existential dissatisfaction with humanity is a key theme throughout Chandler's work, but it is given its fullest expression in *The Little Sister* (1949), in a passage where Marlowe fantasises about evolutionary reversal. Reflecting on the "tired men" he passes on the road, who are headed home to the "whining of their spoiled children and the gabble of their silly wives", he embarks on an embittered soliloquy which ranges through the vapidity and corruption of California ("the department store state", where even the night air is in danger of being optioned) to the tedious and familiar nastiness of his current case.[29] The passage is punctuated with the refrain, "you're not human tonight", and builds towards a melancholic fantasy of discorporation:

> You're not human tonight, Marlowe. Maybe I never was nor ever will be. Maybe I'm an ectoplasm with a private licence. Maybe we all get like this in the cold half-lit world where always the wrong thing happens and never the right.[30]

This moment of existential dejection for the detective–hero is common to all Chandler's novels, but nowhere else is it framed in quite these terms. In wishfully separating himself from the human species and imaginatively embracing a viscous, spectral, formlessness, it is clear that Marlowe sees twentieth-century progress in an extremely negative light; America, from his perspective, is home to an inexhaustible, vacuous consumer and celebrity culture. The passage exemplifies Bird Rose's account of noir as offering us "protagonists whose perspectives uniquely articulate our condition in this dark (Anthropocene) era".[31] Marlowe sees through the relentless commodification of America and wishes to separate himself from "the political, cultural and economic systems [he knows] to be destructive", despite being aware of the impossibility of such a disentanglement.[32] Marlowe's reluctant acknowledgement of human kinship ("Maybe we all get like this") echoes a parallel passage at the end of the earlier novel, *The*

Big Sleep, where he famously declares: "Me, I was part of the nastiness now".[33]

What is also clear, however, is that he does not damn humanity equally; the homosocial sympathy with the "tired men" contrasts with the disdain for the "gabbling wives", denied coherent language here in a way which is consistent with the othering of his monstrous femme fatales and specimens of blonde. What is ultimately at stake in Chandler's work, particularly in these instances of animality, is not a radical or universal existential pessimism, but rather anxiety around the perceived encroachment upon the straight white male's ascendent position, due to rapidly changing gender roles during the depression and post-war years. Marlowe ultimately wants to protect "a world where men live and keep on living".[34] His concern is very much for President Roosevelt's "Forgotten Man"—the intended beneficiary of the New Deal—the hard-working individual who, due to the dire impact of the Depression remains "at the bottom of the economic pyramid".[35] This preoccupation is common to noir more broadly, where, as Megan Abbott argues, "male subjectivity constantly threatens to unravel".[36] Marlowe's musings on discorporation notably come after his agitated consideration of the forceful women in his current case and their unreasonable demands upon him. The fantasy of spectrality then, can be linked particularly to anxieties around masculine identity amidst a changing landscape of gender expectations. It is part of a noir existentialism which has at its core a deep sense of precarity surrounding white, male agency.

Lethem's Evolved Animals

In *Gun, with Occasional Music* (1994) Jonathan Lethem's literalising of Chandler's animal metaphors, picks up on the ethical questions of representing non-humans which Chandler's texts raise. Taking his predecessor's critique of progress to the extreme, Lethem's dystopia is home to evolved animals and "baby heads" (infants whose cognitive development has been accelerated). The evolved animals we meet can be seen as part of Lethem's wider penchant for "calling metaphors' bluffs".[37] In his short story "The Hardened Criminals", for instance, convicts are made into building blocks for prisons and in *Gun, with Occasional Music* the detective, Conrad Metcalf, has been quite literally emasculated by an ex-lover who left before the couple could reverse a temporary operation to exchange genital sensations. The narration repeatedly draws attention to its concretised metaphor conceit, as the following exchange between Metcalf and two aggressive government inquisitors suggests:

> "Angwine's got a problem. His future's all used up. I'd hate to see that happen to a dickface like you".
>
> I turned to Kornfeld, who still hadn't cracked a smile. "Do I have a dick on my face? Tell me honestly".
>
> "You better cancel the fancy punctuation, dickface", said Morgenlander blithely. "Your licence is a piss mark in the snow, as far as I'm concerned". He adjusted his tie, as if his head were expanding and he needed to make some room for it. "Now tell Inquisitor Kornfeld about your trip to the doctor".
>
> "I'm seeing a specialist", I said. "To see if I can have the dick on my face removed".[38]

The typical hard-boiled banter between official law enforcement and maverick PI is combined here with a postmodern self-reflexivity. Metcalf's stubborn and crude literalness in response to his humourless antagonists echoes the text's broader project and there are several related moments throughout the novel where his Marlowesque repartee exhibits a self-conscious concern with language. Marlowe's confident wit has become a form of self-questioning as Metcalf regularly evaluates the effectiveness of his metaphors and admits after one typically self-deprecating and fatalistic exchange with the victim's wife "Even I wasn't sure what that meant" (22). Although it is clearly part of the novel's wider preoccupation with linguistic play and metaphorical slippage, however, the case of concretising animal metaphors raises particular questions about ethics and representation which tie into the text's dystopian concern with the destruction of language and memory that I explore in more detail in the final section.

When interviewed about the response to *Gun, with Occasional Music*, Lethem identifies a source the reviewers had not considered:

> The Ground Zero reading experience for me was *Alice in Wonderland*. When my first novel came out—it was a hard-boiled detective novel that was also a dystopian satire—it was compared to Philip K. Dick and Raymond Chandler, who were almost my two favorite writers in the world. I felt ecstatic. I thought that was fine, they made a great identification, certainly one I could live with. Yet the book is full of talking animals and that's taken for granted, exactly as Lewis Carroll has Alice wonder through Wonderland. We're not concerning ourselves with the *why* of why: the cat, the turtle, the

griffin, or whatever, can act as they do. We're not concerned with why they can speak, but what's wrong in their fictional world: why are they so irritated, why are they so bad? That was the kind of book that I'd written, I'd claimed *Alice in Wonderland* for myself.[39]

Carroll's text, of course offers a model for the linguistic instability which Lethem employs. As Lippit argues: "the fluid relation between physicality and figures of speech places Alice in a world no longer grounded by a referentially stable language. In Wonderland, bodies are allegorised and idioms are literalised in a seemingly unrestricted exchange".[40] Just as with Carroll's novel, we are thrown into Lethem's hard-boiled wonderland with no explanations and have to work to re-orient ourselves. Before we're told of Dr Twostrand's evolutionary therapy which makes "all the animals stand upright and talk", Metcalf rides in the lift with "an evolved sow" "wearing a bonnet and dress but [still smelling] like a barnyard" (18, 12). When first encountering this description there is a temptation to read the "sow" as a derogatory description of a human being, a rural character out of place in the city. Similarly confusing is the "evolved dachshund" who is helped to his feet after Metcalf callously sends him flying through a revolving door (16). As Simons' work suggests, we are so used to animals being used metaphorically to highlight human characteristics or physical attributes that even with the inclusion of the word "evolved" there is a reluctance to leave behind this way of reading animals. Even after the presence of these anthropomorphic animals has been explained, offhand references to the "crabby old goat working a newsstand" can still elicit a double take (39). In these particular instances, crucially, the animals do not speak, which is a key part of the ambiguity.

Animals are exploited and marginalised throughout the novel; sleazy motels do not allow human–animal couples and if they are killed it does not legally count as murder. Animals are second-class citizens and this distinction recalls the racialised demarcations of Chandler's Los Angeles where "shine killing[s]" are regarded as insignificant by the beleaguered LAPD officers who are low enough in the institutional hierarchy to be charged with their investigation.[41] As Abbott argues, in Chandler's cityscape, "the presence of marginalized groups is typically used as a barometer of urban decline".[42] Lethem makes this connection fairly explicitly. In a self-conscious echo of the paranoid, white flight rhetoric of his hard-boiled predecessors, he has Metcalf—a self-confessed bigot—take note of a doorman at one of the more affluent buildings he visits during the course of his investigation: "He was an elderly black human, one of the last you would see in a menial service job. Evolved animals filled pretty much any position they were capable of filling these days, but the Vistamont prided itself on stubborn traditions" (41). Here, we have the same replacement anxiety played out through a human/non-human binary, rather than a racial one. The encroachment of animals into arenas where they are unwelcome is reflected in Metcalf's nostalgic approval for the hotel's "stubborn traditions", which evoke the racialised stratification of Marlowe's LA. On one level then, animals in the novel could be seen as standing (in a fairly heavy-handed extended metaphor) for variously marginalised social groups but, as the novel plays out, the representation of evolved animals resists such a straightforward and reductive codification.

The central animal players fill a variety of familiar hard-boiled roles, from damsel-in-distress sheep, to brother-in-arms ape and, ultimately, kingpin-kangaroo. This cross-species recasting results in some distorted echoes which make explicit the uncomfortable power dynamics of the original sources. For instance, Dulcie, the ewe who knew too much, is introduced in a scene which recalls Marlowe's first encounter with *The Big Sleep*'s slipperless Vivian Sternwood, "whose legs seemed arranged to be stared at".[43] In Lethem's text, however, the detective's attentive gaze is not lascivious (although Dulcie is—it is implied—a sexual being, kept as a mistress by the corrupt Dr Testafer, who eventually murders her): "She kicked the slipper off her right hoof and scratched at her flank in an unnaturally repetitive way, as if she were being bitten by a fly under her wool" (63). Metcalf claims to be unable to read her movements: "The ewe flinched, but it could have meant anything: I was no Doolittle" (62). The illegibility of Chandler's women is here mapped onto species difference and the squeamishness which so frequently accompanies Marlowe's desire is foregrounded in the attention Metcalf bestows upon her creaturely features.

Metcalf does not achieve justice for Dulcie; he lets her killer off in exchange for information, claiming "I'm not such a stickler for animal rights" (237). However, in the world of the novel, where only state-approved narratives are permitted, even ensuring her story is spoken aloud is a rebellious act which serves as a partial retribution for a crime that will not be investigated within the official system of law enforcement. Metcalf is among a vanishing few "to think there was something wrong with the silence that had fallen like a gloved hand onto the bare throat of the city" (230). The state control of history and memory is explicitly framed here as and an act of murderous violence which Metcalf heroically and fatalistically opposes. His desire to confirm who killed the sheep, to include her story in the case's subversive retelling, attests to her significance and points towards his ambivalent view of the evolved animal's status; her death is worthy of an investigation which places him at personal risk.

The callousness suggested in the quip about animal rights is further belied by his visceral reaction in the scene where he discovers her body, earlier in the text:

> I looked at the corpse again, forcing myself to try and find something meaningful in the way the killing had been done, but I couldn't concentrate. The dug out cavity was like a maze that led my eyes again and again to the mutilated black-red heart. I didn't have what it took to search that maze for clues (96).

While alive, Dulcie's body had been subjected to the detective's critical gaze but that detachment vanishes, here. While he is compelled to look, he cannot "concentrate"; he cannot "search". The violence inflicted upon the sheep's body causes a leveling horror. Metcalf's response to Dulcie's mutilation is comparable to his reaction to a human female victim, Celeste Stanhunt, later in narrative where, again, the focus is Metcalf's abject inability to visually process the scene: "I stopped being able to look at the corpse [...] I closed my eyes but the picture didn't go away" (194). These moments reflect back on what is a central feature of the detective genre, particularly in its hard-boiled mode: the fascination with the female corpse. As Glen S. Close argues, hard-boiled fiction "is defined in part by its distinctly detailed description of violently degraded cadavers".[44] Drawing on Julia Kristeva's work, he suggests that the corpse "makes us acutely aware of the precarious borders of our experience as bodies and subjects".[45] Lethem's parallel animal and human death scenes use this abjection to engage with broader questions about the ethics of killing animals.[46] The reduction of both bodies to horrifically mangled flesh, in this case, highlights the particularly "precarious border" between human and non-human animal and Dulcie's death is allowed the same weight as Celeste's. In this way, the text challenges Metcalf's surface disdain for and distinction from the evolved animals; instead we see animals expanding beyond any singular definition as they occupy a variety of roles, including victim, comrade and villain, which inextricably entangle them with their human counterparts.

The Nietzschean Animal and Anthropocene Noir

Throughout the text, we see animals resisting the urge to return to four legs and James Peacock argues that the accelerated evolution of animals in the novel constitutes "a form of willed forgetting [...] a melancholic fixation on the futuristic present, despite the ghostly physical evidence of the 'past' animal body".[47] This taps into a broader philosophical tradition of thinking about animals and memory which can be traced back to a much-quoted section of Nietzsche's parable of the animal. He writes:

> A human being may well ask an animal: "Why do you not speak to me of your happiness but only stand and gaze at me?" The animal would like to answer and say: "The reason is I always forget what I was going to say"—but then he forgot this answer too, and stayed silent: so that the human being was left wondering.[48]

The animal, Nietzsche's parable suggests, exists in the perpetual present; their lack of language is predicated on an amnesiac state, which (for Nietzsche) constitutes a more authentic mode of existence than is possible for humans who are delimited through language and its partial apprehension of truth. As Lippit argues:

> The oblivion of the animal, its inability or refusal to retain language as memory, effects not a world but an affect—an affected world. The animal, for Nietzsche, is "honest" because it lives "unhistorically" [...] By forgetting to speak, erasing history, the animal lives happily, in the present and without secrets [...] What the animal conveys, in Nietzsche's parable, what it discloses, concerns the very possibility of life itself—a world almost without memory.[49]

The novel's plot revolves around a reversal of this Nietzschean distinction of humans and animals, predicated on the ability to retain language through memory. In Lethem's fictional world, which affectionately draws on earlier sci-fi dystopias, radio news is replaced by suggestive music, newspapers consist of captionless pictures and humans further embrace the state-sponsored amnesia by willingly consuming regular doses of Forgettol, a government-distributed drug.[50] It emerges that the victim, a Forgettol addict, accidentally took out a hit on himself, thinking he was arranging the assassination of his wife's lover. After 6 years of being cryogenically frozen as a punishment for his involvement in the case, Metcalf puts together the pieces:

> Maynard Stanhunt was a pretty heavy Forgettol addict—at least by the standards of six years ago. The first time I tried to call him at home, he didn't know who I was. I'd warned Angwine [who was wrongly convicted of the murder] of tangling with people with huge gaps in their day-to-day memory, but I hadn't really thought through the implications myself. Maynard and Celeste were both having affairs in the Bayview Motel, in the same room in fact. With each other [...] The part of him that wasn't getting any action with Celeste was murderously jealous of whoever she was seeing [...] and he told Phoneblum to have his boys blow the guy away. The kangaroo didn't know what Stanhunt looked like [...] He just did as he was told and killed the boyfriend. Stanhunt hired his own hit (250).

This is the fairly standard climactic revelatory moment which detective texts build towards. Following their investigation, the detective assumes control of the disruptive criminal narrative, tying together seemingly disparate pieces of information into a coherent account of the crime. Structurally, most detective texts revolve around the recovery of the past. However, six years after the event, the key human players (blitzed on Forgettol) have no memory of what happened, frustrating Metcalf's desire to achieve justice or closure, which are concepts that rely on continuous narrative. Memory, when Metcalf emerges from the freezer, has been edited, sanitised and outsourced to handheld devices which individuals consult when they need to remind themselves of essential facts of their lives, such as where they work. The text, then, imagines a society where humans have settled into the perpetual present of the Nietzschean animal, although, crucially, this amnesiac state is corrupted, as the details of this present are decided by a totalitarian government. Against this mass forgetting, Metcalf feels "the weight of the past like a ballast, something only I was stupid enough to keep carrying" (234).

The only other character who retains a clear memory of events is the evolved Kangaroo, Joey Castle. Stanhunt was his first kill and in Metcalf's absence he has risen from incompetent gunsel to crime Kingpin. This rise is largely attributable to his continuous grasp of past and present in contrast to his human boss, Phoneblum, who succumbed to the lures of drug-induced amnesia. Joey is such a threatening figure because he retains control of memory and narrative in a world where human identity has been reduced to state-approved soundbites. On one level, in killing Joey, Metcalf attempts to violently reassert the anthropocentric order which has been undermined in the course of his chaotic investigation. However, in choosing to kill the kangaroo rather than the incapacitated Phoneblum, Metcalf acknowledges him as equal. Phoneblum, who cannot remember his crimes is no longer worth his time and cannot provide a satisfactory ending. Instead, he relies on Joey to provide confirmation of his identity as hero; in the generic showdown between criminal and detective Joey is the only one with a sufficient grasp of narrative continuity to fill the role of adversary: "I wanted to kill someone who remembered who I was" (259). The kangaroo is killed because, despite his animal body, he has become too human.

In selecting the kangaroo as his nemesis, Metcalf once again transgresses the social stratifications which place human above animal. Peacock argues that the text's politics are most apparent in Metcalf's desperate (and futile) attempts to restore "a sense of connection, ethical responsibility and collective narrative in the face of atomization".[51] Metcalf seeks kinship with his fellow citizens in very Chandleresque prose where he reflects on the city's nightlife:

> [T]he night was like a dark nullification of the existence of the city. But underneath the night's skirts the city lived on. Disconnected creatures passed through the blackness towards solitary destinations, lonely hotel rooms, appointments with death. Nobody ever stopped to ask them where they were going—no one wanted to know. No one but me, the creature who asked questions, the lowest creature of them all" (130).

Here, distinctions are levelled; the city's inhabitants, whether they are evolved animals, humans or babyheads are all equally wretched "creatures" whose stories Metcalf wishes to understand. This focus on darkness, both literal and metaphorical, of course recalls Chandler's evocation of city streets that are "dark with something more than night"—one of his most quoted phrases which has become an evocative shorthand for noir sensibility.[52] The ominous "something" carries the weight of an oppressive, nullifying existential darkness against which the detective must work. Bird Rose expands on what this darkness entails:

> The noir protagonist is not in control of his or her fate, and there is a widespread dispersal of guilt. The dark is not an aberration but is inseparable from society. And thus the noir sensibility: under the weight of burdens from the past carried into the future as inescapable fate, there is a sensibility of discontent and anxiety, disillusionment and loss of confidence in the possibility of effective agency.[53]

She posits noir, with its sense of dread, shared complicity and its blurring of victims and perpetrators who are inextricably caught up in systems beyond their control, as the most apposite form for expressing our ecological crisis, which is exposing the self-annihilating violence of anthropocentric world views.

As we have seen, however, even as Chandler criticises corrupt power structures, his own version of noir existentialism ultimately reinforces the gendered and racial hierarchies of his time, which have the individualistic white male at their pinnacle. His use of animetaphor is key to this, othering those who threaten to unsettle the status quo; he is writing in the service of a world where "men live and keep on living". Although there is inter-species slippage in his surreal metaphors, the animal imagery is ultimately used to shore up the category of human and to question who should be allowed this status. Lethem's text offers a more radical conception of inter-species entanglements. He uses his hybrid form to expand on the questions opened up by Chandler's work, engaging with issues of memory, language and agency which are central to the detective genre. The self-aware noir pastiche set in a dystopian future allows Lethem to expose the fragility of boundaries between human and non-human animals, extending the ethical reach of the detective's commitments.

Notes

1. Chandler, *Raymond Chandler Speaking*, 74.
2. Qtd in McShane, *The Life of Raymond Chandler*, 64.
3. Tanner, "The Function of Simile", 339.
4. Smith, "The Public Eye", 437.
5. Michael Sorkin, "Explaining Los Angeles", 8.
6. Davis, *City of Quartz*, 18.
7. See, for example: Davis, *City of Quartz*; Fine, *Imagining Los Angeles*; Richard Lehan, "The Los Angeles Novel"; and Scrambray's *Queen Calafia's Paradise*, 11–30.
8. Chandler, *The Big Sleep*, 2; *Farewell, My Lovely*, 119.
9. Lippit, *Electric Animal*, 165.
10. Lippit, *Electric Animal*, 166.
11. Hayward, "Lessons from a Starfish", 260.
12. Chandler, *Playback*, 10.
13. Rose, "Anthropocene Noir", 214–215.
14. I am using metaphor, here, rather than simile. Although most of the examples from Chandler fall under the metaphorical subset of simile, I am look-

ing at them through the context of Lippet's "animetaphor" and the broader figurative use of animals.

15. Effie Rentzou, "Minotaure: On Ethnography and Animals", 32. For a discussion of the transatlantic cross-fertilisation between French surrealism and existentialism and American hard-boiled fiction and film noir see Naramore, "American Film Noir".
16. Chandler, *Farewell, My Lovely*, 7.
17. Chandler, *Farewell, My Lovely*, 149.
18. Chandler, "The Simple Art of Murder", 12.
19. Chandler, "The Simple Art of Murder", 14.
20. Shaw edited *Black Mask*, the pulp magazine which launched the careers of Chandler and Hammett, among many others. In an article for *The Writer's Digest*, he argued that the writer's key task is to have "his characters move and talk, act and react as real human beings would do in like situation" ("Do you Want to be a Writer or do you Want to Make Money?").
21. McCann, "The Hardboiled Novel", 44.
22. Abbott and Lethem, "It's High Time you Read The Big Sleep".
23. Chandler, *The Big Sleep*, 12; 171.
24. Chandler, *The Big Sleep*, 238.
25. Olson, *Criminals as Animals*, 1.
26. Simons, *Animal Rights*, 87.
27. Abbott, "The Big Sleep"; Chandler, *The Long Goodbye*, 76.
28. A similar use of animality to other is frequently deployed when Marlowe observes racial difference. In *Farewell, my Lovely*, for instance, the villain's "Indian" driver is described as having the "apparently awkward legs of a chimpanzee" and "the earthy smell of primitive man", adding to the early twentieth-century tradition of racist Darwinian discourse which places whiteness as the pinnacle of human evolution (124).
29. Chandler, *The Little Sister*, 78–80.
30. Chandler, *The Little Sister*, 80.
31. Rose, "Anthropocene Noir", 214.
32. Rose, "Anthropocene Noir", 215.
33. Chandler, *The Big Sleep*, 250–251.
34. Chandler, *Farewell, My Lovely*, 165.
35. Roosevelt, "The 'Forgotten Man'".
36. Abbott, *The Street Was Mine*, 27.
37. Silverblatt, "An Interview with Jonathan Lethem", 41.
38. Lethem, *Gun, with Occasional Music*, 35. Subsequent quotations will be cited in parenthesis.
39. "Jonathan Lethem by Betsy Sussler".
40. Lippit, *Electric Animal*, 137–138.
41. Chandler, *Farewell, My Lovely*, 18.
42. Abbott, *The Street Was Mine*, 106.

43. Chandler, *The Big Sleep*, 16.
44. Close, *Female Corpses in Crime Fiction*, 9.
45. Close, *Female Corpses in Crime Fiction*, 14.
46. Lethem also explores this issue in the short story "Pending Vegan", where a depressed father takes his daughters to SeaWorld and is unsettled by the nightmarish juxtaposition of the children's love for the captive animals and their unquestioning consumption of "the huge cartilagionous drumstick" he purchases from one of the theme park's food outlets (153).
47. Peacock, *Jonathan Lethem*, 27.
48. Nietzsche, "History for Life", 60–61.
49. Lippit, "Magnetic Animal", 1116.
50. Lethem lists J.G. Ballard, George Orwell, Philip K. Dick, Aldous Huxley and the Brothers Strugatsky among his influences for *Gun*. Lethem, "The Art of Fiction", 57.
51. Peacock, *Jonathan Lethem*, 29.
52. Chandler, "Introduction to 'The Simple Art of Murder'", 1016.
53. Rose, "Anthropocene Noir", 215.

Works Cited

Abbott, Megan E. *The Street Was Mine: White Masculinity in Hardboiled Fiction and Film Noir.* New York: Palgrave, 2002.

Abbott, Megan E. "The Big Seep: Reading Raymond Chandler in the age of #MeToo". *Slate.* 9th July, 2018. https://slate.com/culture/2018/07/raymond-chandler-in-the-age-of-metoo.html

Abbott, Megan and Jonathan Lethem, "It's High Time you Read *The Big Sleep*". *GQ*, 17 July 2018. https://www.gq.com/story/its-high-time-you-read-the-big-sleep

Deborah Bird Rose. "Anthropocene Noir". *Arena Journal* no.41/42. (2013/2014). 206–219.

Chandler, Raymond. *Playback.* London: Penguin, 1961.

Chandler, Raymond. *Farewell My Lovely.* London: Penguin, 1973.

Chandler, Raymond. "The Simple Art of Murder". In *The Simple Art of Murder.* New York: Ballantine, 1972.

Chandler, Raymond. "Introduction to "The Simple Art of Murder". Rpt. in *Later Novels and Other Essays*, 1016–1019. New York Library of America, 1995.

Chandler, Raymond. *Raymond Chandler Speaking*, eds. Dorothy Gardiner and Katherine Sorley Walker. Oakland: University of California Press, 1997.

Chandler, Raymond. *The Big Sleep.* London: Penguin, 2005.

Davis, Mike. *City of Quartz.* New York: Vintage Books, 1992a.

Fine, David. *Imagining Los Angeles: A City in Fiction.* Albuquerque: University of New Mexico Press, 2000.

Hayward, Eva. "Lessons from a Starfish". In *Queering the Non/human*, eds. Myra Hird and Noreen Giffney, 248–264. Aldershot: Ashgate, 2008.

"Jonathan Lethem by Betsy Sussler". *BOMB Magazine*, 3 Oct, 2008. https://bombmagazine.org/articles/jonathan-lethem/

Lehan, Richard "The Los Angeles Novel and the Idea of the West". In *Los Angeles in Fiction: A Collection of Essays*, ed. David Fine, 29–42. Albuquqrque: University of New Mexico Press, 1995.

Lethem, Jonathan. "The Hardened Criminals". In *The Wall of the Sky, the Wall of the Eye*. 172–210. London: Faber, 2014.

Lethem, Jonathan. "Pending Vegan". In *Lucky Alan and Other Stories*. 137–157. London: Penguin, 2015.

Lethem, Jonathan. "The Art of Fiction No. 177: Jonathan Lethem". In *Conversations with Jonathan Lethem*, ed. Jaime Clarke, 53–70. Jackson: University of Mississippi Press, 2011.

Lippit, Akira Mizuta. *Electric Animal: Towards a Rhetoric of Wildlife*. Minneapolis: University of Minnesota Press, 2000.

Lippit, Akira Mizuta. "Magnetic Animal: Derrida, Wildlife, Animetaphor". *MLN* 113 no 5 (1998). 1111–25.

MacShane, Frank. *The Life of Raymond Chandler*. London: Jonathan Cape, 1976.

Davis, Mike. *City of Quartz: Excavating the Future in Los Angeles*. London: Vintage, 1992b.

McCann, Sean. "The Hardboiled Novel". In *The Cambridge Companion to American Crime Fiction*, 42–57. Cambridge: Camridge University Press, 2010.

Naramore, James. "American Film Noir: The History of an Idea". *Film Quarterly* 49.2 (1995–1996), 12–28.

Nietzsche, Frederick. "On the Use and Disadvantage of History for Life". In *Untimely Meditations*, ed. Daniel Breazeale, translated by R.J. Hollingdale, 57-124. Cambridge: Cambridge University Press, 1997.

Olson, Greta. *Criminals As Animals From Shakespeare to Lombroso*. Berlin: De Gruyter, 2013.

Peacock, James. *Jonathan Lethem*. Manchester: Manchester University Press, 2012.

Rentzou, Effie "*Minotaure*: On Ethnography and Animals", *Symposium: A Quarterly Journal in Modern Literatures*, 67 no 1 (2013). 25–37.

Roosevelt, Franklin D. "Radio Address From Albany, New York: The 'Forgotten Man' Speech". *The American Presidency Project*. <https://www.presidency.ucsb.edu/node/288092>

Scrambray, Kenneth *Queen Calafia's Paradise*. Madison: Farleigh Dickinson University Press, 2007.

Shaw, Joseph T. "Do you Want to be a Writer or do you Want to Make Money?". *The Writer's Digest*. May 1934. Reprinted Online at *Black Mask Magazine*: <https://blackmaskmagazine.com/blog/do-you-want-to-become-a-writer-or-do-you-want-to-make-money-by-joseph-t-shaw/>

Simons, John. *Animal Rights and the Politics of Literary Representation*. New York: Palgrave, 2002.
Silverblatt, Michael. "An Interview with Jonathan Lethem" in *Conversations with Jonathan Lethem*, ed. Jaime Clarke, 37–42. Jackson: University of Mississippi Press, 2011.
Sorkin, Michael. "Explaining Los Angeles" in *California Counterpoint: New West Coast Architecture, 1982*. New York: Rizzoli, 1982.

Index[1]

[1] Note: Page numbers followed by ‘n’ refer to notes

R. Hawthorn, J. Miller (eds.), *Animals in Detective Fiction*, Palgrave Studies in Animals and Literature,
https://doi.org/10.1007/978-3-031-09241-1

R

S

Y

Printed in the United States
by Baker & Taylor Publisher Services